STP 1254

Science and Technology of Building Seals, Sealants, Glazing, and Waterproofing: 3rd Volume

James C. Myers, Editor

ASTM Publication Code Number (PCN):
04-012540-10

ASTM
1916 Race Street
Philadelphia, PA 19103
Printed in the U.S.A.

ISBN: 0-8031-1993-3
PCN: 04-012540-10

Peer Review Policy

Each paper published in this volume was evaluated by three peer reviewers. The authors addressed all of the reviewers' comments to the satisfaction of both the technical editor(s) and the ASTM Committee on Publications.

The quality of the papers in this publication reflects not only the obvious efforts of the authors and the technical editor(s), but also the work of these peer reviewers. The ASTM Committee on Publications acknowledges with appreciation their dedication and contribution to time and effort on behalf of ASTM.

Printed in Scranton, PA
June 1994

Foreword

The Third Symposium on Science and Technology of Building Seals and Sealants was held in January 1993 in Fort Myers, Florida. The symposium was sponsored by ASTM Committee C-24 on Building Seals and Sealants. Papers deal with the fields of building seals, sealants, glazing, and waterproofing. James C. Myers of Simpson Gumpertz & Heger, Inc., Arlington, MA, served as the editor for this publication.

Contents

Overview

Technology of building seals and sealants continues to develop and change. A host of new products and applications have evolved since the first elastomeric sealants in the 1950's, and new sealant technologies are on the horizon. This on-going development has produced some remarkable advances and has kept ASTM Committee C-24 on Building Seals and Sealants busy over the last 34 years as we developed over 70 sealant-related standards.

ASTM C-24 has presented symposia, with resulting ASTM Special Technical Publications (STP), yearly since 1988, to help disseminate current knowledge on all aspects of building seals, and plans to continue this work. This publication contains eight peer-reviewed papers on sealant durability, design, and testing topics.

Durability

Lacasse's paper reviews past research and methods used to assess long-term sealant performance and discusses future research needed to develop test methods. Leonard provides new research data on urethane reversion problems, to add to the pool of data presented in the 1991 and 1992 symposia. Beech continues his research into sealant cure and durability on which he presented his initial findings in 1992. Fiorillo reviews the track record of some polysulfide installations.

Design

Rutila provides some guidelines for designing expansion joint seals in deck waterproofing, to expand on his 1989 paper on building deck waterproofing. Lacasse describes an expert system that he is developing to aid designers in selecting sealants.

Testing

Al-Qadi examines a new laboratory test method to simulate field movement of joints. Wolf describes some new laboratory techniques to assess adhesion to aluminum.

Hopefully, these papers help answer some questions that have arisen in sealant technology recently, provide the foundation for development of new standards, and stimulate further research. ASTM C-24 welcomes your comments and suggestions regarding our symposia program. I thank the authors, peer reviewers, and ASTM staff whose generous contribution made this publication possible.

James C. Myers, Symposium Chairman and Editor
Simpson Gumpertz & Heger, Inc.
297 Broadway
Arlington, Massachusetts 02174

Durability

Michael A. Lacasse[1]

Advances in Test Methods to Assess the Long-Term Performance of Sealants

REFERENCE: Lacasse, M. A., "**Advances in Test Methods to Assess the Long-Term Performance of Sealants,**" *Science and Technology of Building Seals, Sealants, Glazing, and Waterproofing: 3rd Volume, ASTM STP 1254,* American Society for Testing and Materials, Philadelphia, PA, 1994, pp. 5–20.

ABSTRACT: In this review, a number of aspects related to sealant testing and evaluation will be addressed. Those factors that are most important in determining the useful life of sealants, as well as the choice of the most appropriate test methods to assess their performance are discussed. Some of the topics that are treated include, for example, studies undertaken to determine the amount and rate at which building joints move and their significance in relation to the effect of movement rate on the mechanical properties of sealants; the relevance of the failure envelope in evaluating the performance of sealants; the effect of daily cycling on the extended sealant; the use of strain-cycling movement when predicting sealant performance; the significance of light and moisture on the overall degradation of the sealant; the importance of movement in the deterioration of sealants joints. The nature of future research regarding the long-term performance of sealants is identified.

KEYWORDS: sealant, sealant evaluation, sealant performance, joint movement, mechanical properties of sealants, test methods sealant test apparatus, degradation of sealants

The outside envelope of a building undergoes cyclic movements as it responds to changing temperature and moisture conditions, the movement being accommodated at the joints. To prevent the passage of water, air, or dust, the joints are sealed with elastomeric based compounds called sealants. These materials are applied as liquids against a back-up material and harden through chemical or physical means to form rubber-like compounds capable of withstanding repeated cycles of extension and compression. In general, the sealant is extended at low temperature and expand with increasing temperature.

The demands on sealants are severe in climates such as those that exist in Canada because of the large temperature fluctuations on the outside of buildings. To determine the performance of sealants, test methods are required that provide an indication of their performance. To date, there is no standard laboratory test applicable to sealants that is derived directly from behavior on outdoor exposure. The existing CGSB and ASTM tests subject a sealant to various mechanical, temperature and immersion cycles that are not based on exterior exposure of sealants or observations of building joints. Furthermore, the existing tests call for curing of sealants and this, no matter how carefully chosen, does not reproduce the curing in a building joint. Even if the extent of movement is small, it is significant because previous investigations demonstrated that it is the most important factor influencing performance for silicone [*1*], polysulfide [*2*], and polyurethane sealants [*3*]. More recently, it has been further demonstrated that cyclic

[1] Research Associate, Building Science Engineer, National Research Council of Canada, Institute for Research in Construction, Materials Laboratory, Ottawa, Ontario, Canada K1A OR6.

movement during sealant cure has a significant affect on the mechanical properties of these products [*4*].

Other reviews have focused on establishing the state-of-the art of assessing the long-term durability of sealants [*5,6*]. Beech has provided an extended examination of the many different methods used to assess the movement capability of building sealants [*5*]. From this review it was apparent that the degree of cure of sealant specimen prior to assessing their performance was considered of particular importance since it is known that the degree of cure at any given time is also a measure of the extent to which the network structure has developed. Consequently similar sealants must be cured to the same extent in order that meaningful comparisons can be made. This suggests that a particular cure regime be established such that all sealants that cure under similar mechanisms can be evaluated on a equal basis. The other alternative is that sealants be evaluated as they cure, that is, they should undergo cycles of movements as soon as that are cast, as is known to occur in building joints. Beech [*5*] suggests that testing of slow curing sealants (that is, moisture cured sealants) be done on samples which have sufficiently cured. The periods of cure for such sealants would be 3 to 12 months or until stable properties have been achieved.

Another recommendation that was put forward was the necessity for attempting to collect and assess data related to determining the aging behavior of sealants. This was undertaken by Wolf [*6*] in a comprehensive study of factors causing degradation in building joint sealants. From this review, it was possible, by comparing the results acquired using different test methods, to demonstrate the inherent deficiencies of the accelerated aging methods currently in use, and to provide some guidance as to the type of studies which could be useful in establishing the long-term durability of sealants. In this regard, Wolf clearly states that cyclic movement is the preponderant factor in the aging of joint sealants. Suggestions for future work relate to the incorporation of cyclic movement to aging factors, as well as arriving at an appreciation of the effect of various sources of ultraviolet radiation on the degradation of sealants.

The focus of this brief review is to draw attention to the work undertaken over more than two decades at the laboratories of the Institute for Research in Construction (formerly the Division of Building Research) in the area of sealant life prediction. Particular emphasis will be given to the more recent contributions that relate to correlating the performance of sealants subjected to laboratory fatigue testing with those subjected to testing on a strain cycling exposure rack. Hence, this paper is to review some of the previous work in relation to the current research related to establishing suitable tests to determine the long-term performance of sealants, to briefly describe some of the more recent advances, and finally, to outline areas for future research.

Performance Testing of Sealants

Joint Movement and Sealant Characterization

Ashton has outlined some of the more important factors that should be taken into consideration when developing predictive tests for assessing the durability of building materials [*7*]. These are provided below with appropriate explanations:

i) knowledge of the fundamental processes causing degradation;
ii) the nature of the test conditions in relation to service conditions;
iii) the transferability of tests developed for a particular material to another material;

A knowledge of the degradative processes that occur in materials is particularly important. A reliable accelerated test can only be developed based on an understanding that causes deterioration in a given material. Furthermore, acceleration of degradative factors must relate to service conditions to obtain meaningful results. Hence, a reliable means of accelerated testing is achieved by continuity of exposure to the appropriate factors at their maximum levels rather than by the use of excessive levels. This implies that the conditions to which a material is exposed in practice should be measured to provide a valid basis for the test. The more the test conditions depart from the natural levels, the more restricted the applicability of the test will be. Finally, tests developed for one material should not be applied to another without first establishing that the properties of the second are related to the basis for the test.

These principles are particularly important because they form the basis for establishing a useful performance test and this same process was adopted by Karpati in the development of a performance test to assess the durability of sealants.

In the case of joint sealants, it was recognized from the outset of the research program that the tests that come closest to performance testing of sealants are those that subject the sealant to cyclic movement and simultaneous temperature change [*8*]. It was, however, considered a complex test because in order to be successfully completed, it required a knowledge of the fundamental properties of the materials being tested as well as an understanding of the rates and amplitude of movement and the associated temperature range in which the materials are required to perform in the field. Hence initially, two avenues of research were simultaneously pursued to provide the required information: basic information related to the mechanical properties of the polymer based materials used in sealants was obtained by subjecting model sealant beads to tensile tests at various temperatures and strain rates and studies were undertaken to assess the nature of joint movement that occurs on the outside of a building.

The studies related to joint movement showed that joint between adjacent building elements correlated reasonably well with changes in surface temperature of the cladding material [*9*]. Furthermore, it was shown that the movements imposed on a sealant consist of daily cycles superimposed on yearly cycles. For example, in the case of a joint that moves ±25% of its width in a year, the daily movement would be, on the average, ±4% of joint width.

Another important observation obtained from these initial studies was that the rates of movement in the joints were comparatively small in relation to the rates at which sealants are commonly tested. This particular information was useful in adjusting the rate at which tensile tests were conducted such that meaningful properties could be obtained.

Failure envelopes are often used to characterize the nature of failure in polymeric based materials and are plotted as log strain against log stress. This was also done for the results obtained from tensile tests, however, it became evident that in the case of sealants, it is more relevant to plot results as log strain against log time when evaluating the performance of sealants because what is important is the extendibility of the sealant at a given time period. Furthermore it was determined, after having conducted a series of cyclic tests [*10*], that the results of these tests could also be plotted in the same manner as that used for tensile tests, thereby establishing a link between cyclic and tensile testing. It also provided a useful means of being able to interpret data from specimens placed on exposure racks and exposed to different levels of strain. What appeared complex in the initial research stages now became the basis for assessing the performance of sealants. What follows is a review of the methods utilized to assess the performance of sealants with cyclic movement as the main factor causing degradation.

Review of Methods Using Cyclic Movement

The use of cyclic devices to subject sealants to the type of movements that occur on the outside of the building envelope have been achieved in numerous ways and for the purposes of this review, have been categorized into three distinct classes: natural weathering exposure rigs, manually adjusted vises and fully automated mechanical devices. Most natural exposure rigs were developed by the different research organizations as a tool to enable a more pertinent study of sealant durability whereas considerable monies have been expended by sealant manufacturers on the development of, in certain cases, very elaborate testing devices. An exhaustive review of the results from tests using the former test methods has already been reported [*6*] and will not be repeated here, although some of the more salient aspects of the results obtained from the use of the strain cycling exposure rack (SCR) at the IRC will be outlined. With respect to the test methods relating to mechanical devices, mention is made of the review provided by Cook [*11*] on such type of automated devices. Unfortunately, but understandably, very little information has been published on the results of fatigue tests on sealants from the use of automated devices. A brief outline of the few results that are available is provided in this section.

Natural Weathering Exposure Rigs—Brown [*12*] reports that the initial description of cyclic movement testers used in his study on joint sealants was provided in the papers of Ballantyne [*13*] and Ballantyne and Brown [*14*]. In these devices, the difference in thermal coefficients of expansion of timber and aluminum was used to impart the cyclic movement to a sealant specimen in shear. A diagram of the apparatus is shown in Fig. 1. The chief use of these durability tests was to provide some indication of the relative performance of the various types of sealants being evaluated. The devices were used to evaluate sealant specimens over a two-

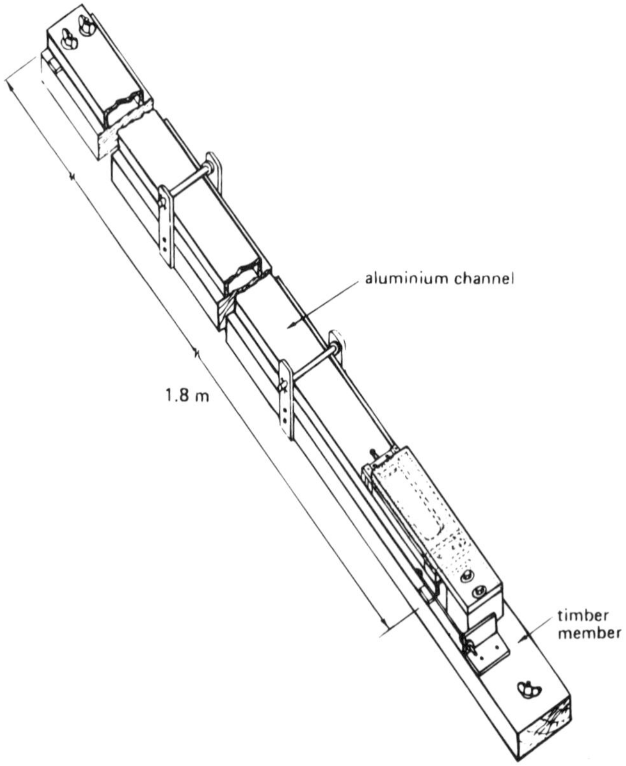

FIG. 1—*CSIRO cyclic movement tester* [*12*].

year exposure period in Highett, Australia. The type of failure as well as the time of failure are reported for the various classes of sealants being tested, which included: oil-based glazing compounds, butyl-based mastics and a number of elastomeric-based sealants.

This same device was used by researchers at the BRANZ [*15,16*] (Building Research Association of New Zealand) to evaluate the durability of a range of different types of sealants including single component silicones, latex and solvent based acrylic products as well as single and multiple component polyurethane and polysulfide-based products. Results from sealants subjected to natural exposure conditions (Judgeford, Wellington, N.Z.) for an initial period of 24 months [*15*] and thereafter a report after six years of natural weathering [*16*] indicated that the device was useful in determining the relative durability of the various types of sealants, producing the types of failures that occur in practice. It was further noted that exposure of the sealants without cyclic movement was not sufficient to reliably evaluate sealant durability.

Burstrom reports on the use of a natural weathering exposure rig which uses the differences in coefficients of thermal expansion of a relatively massive (25 × 900 × 6000 mm) concrete element and a pair of aluminum tubes running along the length of the element [*17*]. In this study, polysulfide, polyurethane, acrylic, as well as an oleo-resinous product, were evaluated on the basis of changes in mechanical properties in tension after various periods of exposure. It is not readily apparent from a review of this article the extent to which specimens were exposed, hence an evaluation of the usefulness of the natural weathering rig is not possible.

The Princes Risborough Laboratory (PRL), Building Research Establishment, UK, has devised a natural exposure movement simulation rig which uses black PVC driving bars to respond to temperature change and give rise to movement at a joint. The simulated sealant-filled joints exposed to weather on these rigs show daily and seasonal movement similar to those on buildings [*18*]. Reports on the usefulness of this device have yet to be published.

At the IRC, a weathering rack for sealants (Fig. 2) was developed to provide a test facility that could simultaneously subject sealant specimens to cyclic movement and temperature change, similar to that known to occur in various types of joints in buildings [*19*]. The

FIG. 2—*Strain-cycling exposure rack* [*19*].

movement is produced by using the differential movement between steel and aluminum. The rack can accommodate 216 specimens with imposed movements ranging from ±9% to ±30% joint width in three-quarters of the rack and the remaining 54 specimens move ±19%. With this arrangement, movement above which a sealant starts to fail can be determined for a series of specimens exposed to various amounts of yearly movement. Those specimens that undergo identical movement can be subjected to repeated tests at given time intervals.

Results of tests on one-part chemically cured silicone sealants using this device indicate that strain-cycling movement is the predominant factor which causes failure during weathering of these sealants; outdoor weathering alone is negligible [1]. Failure is also influenced by the extent of movement, becoming more extensive at higher yearly strain cycles. Tensile tests reveal a notable decrease in strength beyond about ±22% movement.

In a similar study, a two-part polysulfide sealant formed cavities on the side sheltered from direct sunshine [2]. The same material, when weathered without cycling or stored in the laboratory, did not undergo permanent deformation. This underlines the importance of the effect of cyclic movement on the performance of two-part polysulfide sealants, revealing it as the most important factor in test methods intended to evaluate their performance.

The SCR was also used to assess the performance of low movement capability sealants that included various latex and solvent-borne acrylic sealant products and a chlorosulphonated polyethylene sealant [20]. In this study, the use of the rack provided a quick assessment of the movement capability of the sealant because failures started to occur after only a few months of exposure.

A procedure was also developed for the rapid evaluation of silicone sealants in which unsatisfactory products could be identified within a two-month period [21].

Finally, the rack was used to assess the performance of various types of polyurethane based sealants [3]. One single and three multi-component polyurethanes and one epoxy-polyurethane sealant were evaluated. Again, the results for these tests indicate, as was shown in other studies, that joint movement is the predominant factor in the failure of these products.

Mechanical Devices—A manually operated device (Fig. 3) was designed to enable individual sealant specimens to be subjected to cyclical movement [22]. Hence this inexpensive device,

FIG. 3—*Cyclic movement device* [22].

essentially developed as a means to imitate that which occurs on a strain cycling rack (SCR), can subject sealant specimens to extension and compression while exposed outdoors or to various artificially produced conditions [*23*]. Whereas in the case of the SCR, joint movement is continuous and automatic, this device must be adjusted manually to provide the necessary changes in joint width.

A number of studies were completed using this device [*24–26*] and in the case of the work carried out by Karpati [*24,25*], the results suggest that the laboratory test method, devised essentially as a fatigue test, can be correlated to those results obtained from sealant specimens exposed on a strain cycling exposure rack.

Initial work using this device was used to assess the performance of two-part polysulfide sealants which had been previously tested on the SCR [*2*]. In an effort to duplicate practice as closely as possible, newly cast specimens were exposed outdoors as soon as they could be handled. The vise widths were changed once a month, each change being one third of the total for the year, the total change ranging from ±10% to ±80% (Fig. 4). Based on the results of these trials, it was recognized that the polysulfide sealant underwent continuous change in appearance which, upon further examination, could be categorized into three distinct stages of permanent deformation. The first stage represents the largest joint width movement (± %) at and below which permanent deformation is negligible; the second stage is characterized by the formation of lengthwise bulges in sealant specimens; the third stage provides an indication of the smallest movement (± joint width change, %) at and above which holes develop (Fig. 5). The observed stages of permanent deformation are shown as a function of the number of winters of exposure. Stages one and two of specimens exposed on the SCR from the previous study [*2*], are represented by a sinusoidal line marked SCR1 and SCR2 at the approximate time in which they occurred. The stages of observable deformation develop in these sealant specimens at lower movements because these undergo diurnal movements on the SCR that is

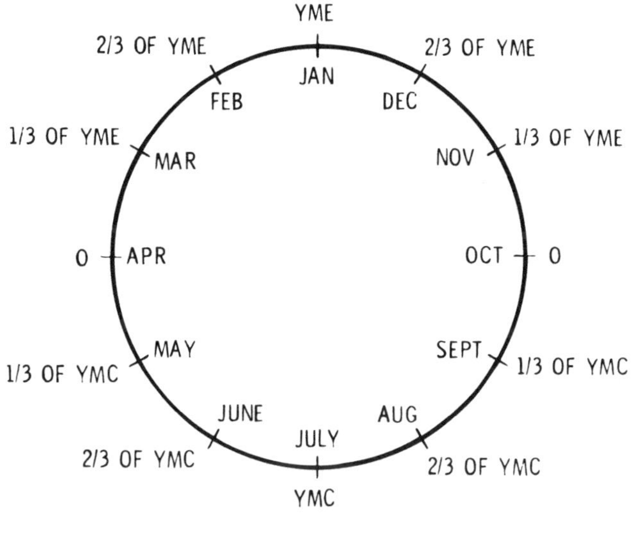

FIG. 4—*Time schedule of joint width changes* [*24*].

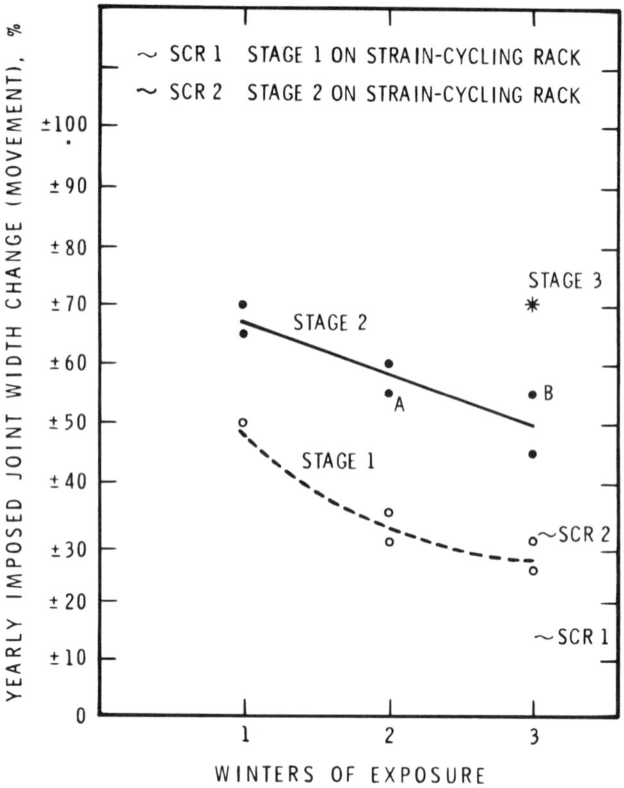

FIG. 5—*Stages of permanent deformation* [*24*].

in the order of one-fifth the yearly movement and hence, the same degree of cumulative damage occurs for a given time, at lower annual movements.

Based on these initial studies, it was concluded that it was possible to produce with hand-operated vises the same permanent deformation of sealant specimens that occurs on the SCR. Another observation was that any elastomeric based sealant could be evaluated using this technique provided the processes of permanent deformation were similar to those observed to occur in the two-part polysulfide sealant.

It was apparent that on the basis of previous weathering tests, the most important factor leading to permanent deformation and eventual failure of the joint was cyclic movement. Hence, it was useful to know whether failure produced by cycling during natural weathering would also result solely from cycling in the laboratory [*25*].

Sealant specimens of the same two-part polysulfide sealant as in the previous studies [*2,24*] were mounted after accelerated curing (4 d at 30°C), on vises and extended and compressed to different amplitudes (from ±10 to ±80%) and at three cycling rates (1, 2.5 and 7.5 cycles/week). Results from these tests presented in Fig. 6, indicated that all three stages of permanent deformation, previously observed on specimens mounted on vises and exposed outdoors on a SCR, were produced using this laboratory test method. Furthermore, the results of any of the stages, including stage three (i.e. sealant perforation) could be plotted in such a manner as to provide an indication of the long-term performance of the sealant. For example, the best-fit lines of stage three of the laboratory fatigue tests give movements of ±19% when extrapolated

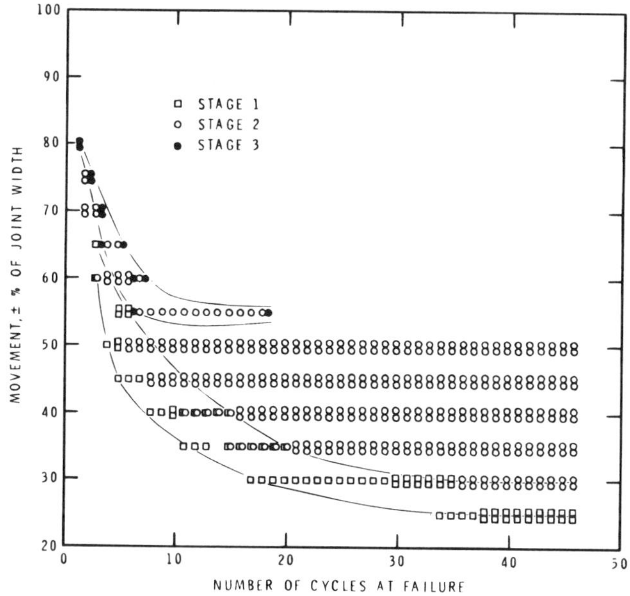

FIG. 6—*The three stages occurring at 7 1/2 cycles per week* [*25*].

to 20 years (Fig. 7). It is to be noted that the results obtained from these fatigue tests are consistent with those obtained from sealants subjected to outdoor exposure and cyclical movement, the movement being achieved either through the use of vises or the SCR (Fig. 8).

It remains to be determined, however, if the degree of cure of specimens used in this experiment would necessarily change the extent to which results can be correlated with that of work carried out in the outdoor exposure tests.

Lacasse [*26*] used hand-operated vises similar to that devised by Karpati with the exception that up to six specimens could simultaneously be subjected to joint movements. Thirty-six testing devices, of the type shown in Fig. 9, were used to subject two commercially available polyurethane based sealants, cured for two months in the laboratory, to a cyclic test program comparable to that undertaken by Karpati in the laboratory fatigue test of a two-part polysulfide sealant [*25*]. The sealants in this study were subjected to 7.5 cycles per week in amplitudes of ±25 to ±80% of the joint width in steps of 5%. Triplicate specimens were used at each amplitude. Figure 10 depicts results of sealant specimens at the various stages of deformation, described earlier. The progressive deterioration of the sealant bead is clearly noticeable for specimens A2 subjected to ±70% width change after 2, 3, 4 and 4.5 cycles respectively. A more gradual degeneration is observed at smaller strain amplitudes as shown in the subsequent sequence of photographs depicting type A3 sealants undergoing movements of ±40, 50 and 60% respectively. Based on the results of this test and using the same technique to predict performance as that used by Karpati [*25*], a twenty-year life expectancy (to stage 3) is achieved for the A3 sealant if the strain program does not exceed the ±15% joint width range. It must be stated that these results have not been correlated to long-term outdoor exposure tests, as was the case in the aforementioned studies by Karpati. This would necessarily be required in order to validate the prediction of durability resulting from the interpolation of data acquired from the fatigue test.

Another aspect that has practical implications is the possibility of increasing the test rate without unduly affecting the nature of the results. This is particularly relevant in the development

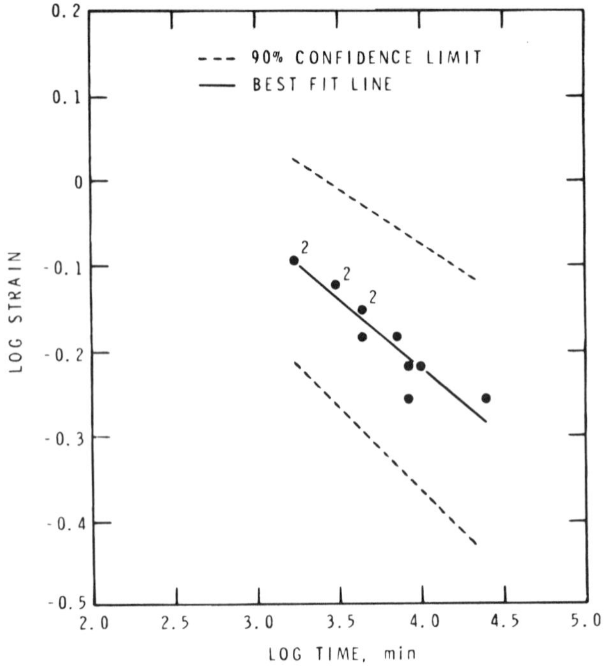

FIG. 7—*Stage three results from 7 1/2 cycles per week* [*25*].

of an automated fatigue testing device capable of subjecting sealants to much higher rates of cyclic deformation than can be achieved using manually adjusted vises. It is evident that there exists a potential to reduce the amount of time required to assess the performance of sealants in using such a device.

Many mechanized testing rigs have been devised over the years and, as mentioned earlier, Cook [*11*] has described a number of different types of devices built by testing agencies, manufacturers of testing equipment or again, sealant producers. More recently, the BRE has developed an apparatus for testing the movement capability of building sealants. It is capable of evaluating up to 40 specimens (BSI standard with a 50 × 12 × 12 mm bead) in compression or extension by 50%. An environmental chamber houses the test rig and can be maintained at temperatures ranging from −20 and +70°C. The entire testing mechanism is computer controlled and the test temperature is adjusted according to the direction of travel, whereby the chamber is cooled when the specimens are being extended and heated when specimens are being compressed. Results on the use of this device have yet to be published.

In fact, few results are available on the cyclic fatigue of building sealants. Matsumuto [*28*] does however, report on the use of a cyclic test device complete with environmental chamber (−10 to 100°C) capable of subjecting up to 22 sealant specimens to a cyclic program of 300 cycles per minute. The report suggests that sealants can readily be classified on the basis of their fatigue resistance, evaluated on the basis of the resilient energy (toughness) of the sealant before and after being subjected to cyclic movement. Although this method is reportedly capable of ranking the sealants in terms of their cyclic durability, it does not provide an indication as to how long sealant materials might expect to remain functional in practice. No further work using this device has been reported.

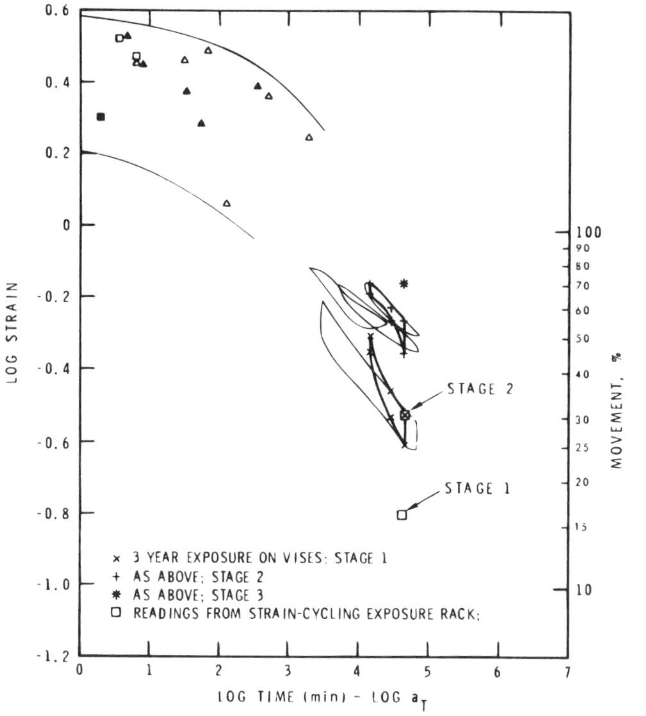

FIG. 8—*Tensile and laboratory fatigue test results* [25].

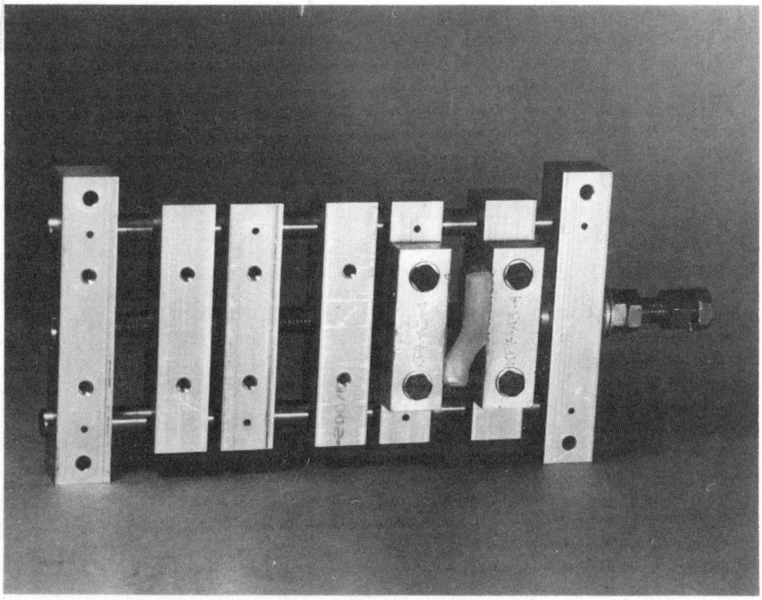

FIG. 9—*Cyclic movement device capable of accommodating up to six sealant specimens* [26].

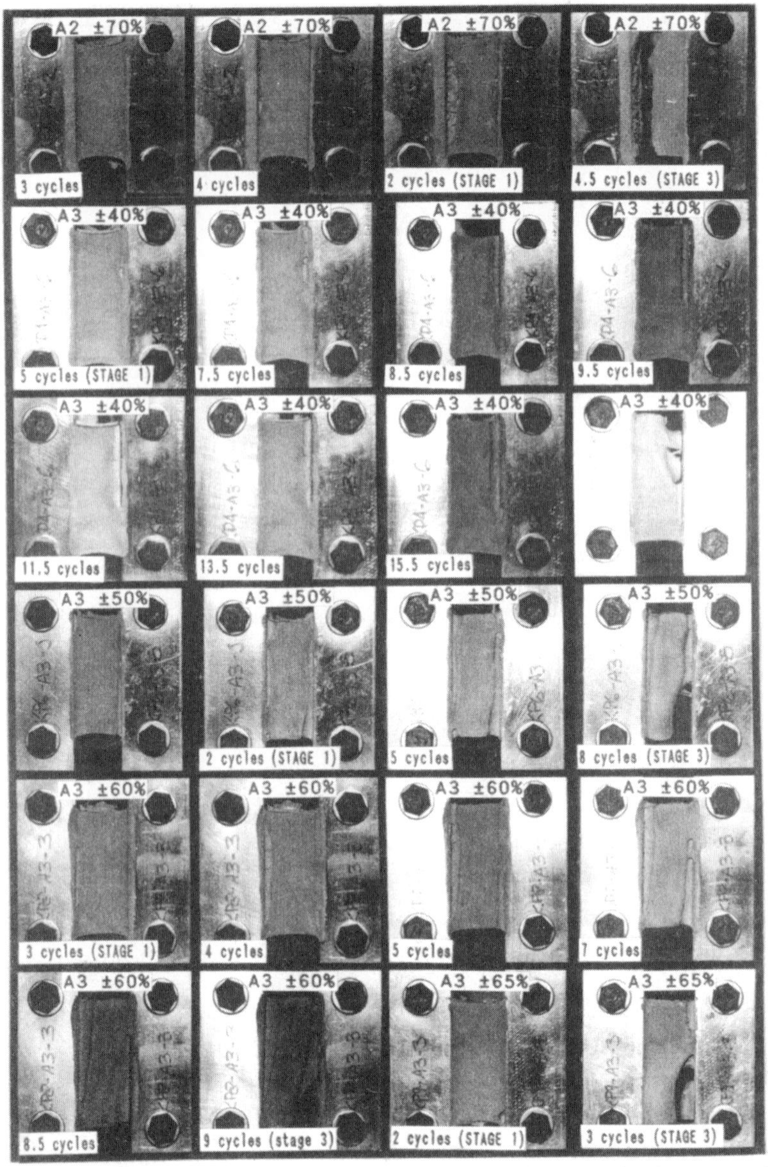

FIG. 10—*Composite photograph showing various stages of deformation for three types of polyurethane based sealants* [*26*].

Lacasse [*26*] produced a similar device as that developed at the BRE in that up to 36 sealant specimens could be subjected to simultaneous cyclic movement in an environmental chamber. The testing frame and hydraulic press were part of an existing test platform otherwise used for testing concrete cylinders, but in this case fitted with a sealant test rig, as shown in Fig. 11. The press was connected to a hydraulic power unit as well as the necessary control apparatus and central processing unit (CPU), such that a given extension-compression program could be set up directly from the CPU. This testing system was used to assess the cyclic

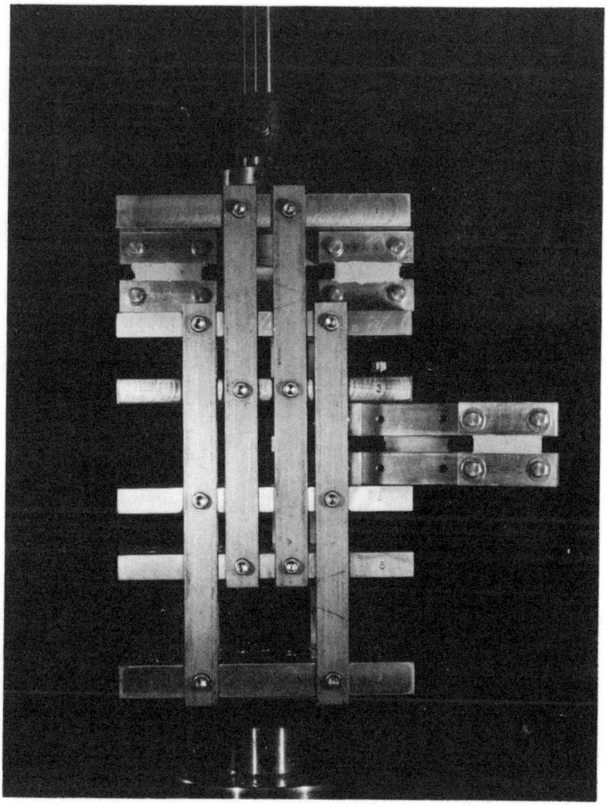

FIG. 11—*Thirty-six specimen sealant test rig* [26].

performance of various polyurethane-based elastomer formulations and subjected replicates of six specimens for each polyurethane type to a series of five strain amplitudes ranging from ±25 to ±60% of the joint width. The device was useful in being able to rank the performance of the various sealants in terms of their resistance to cyclic fatigue. A similar device is currently being developed at the IRC for eventual use in the study of the fatigue resistance and long-term performance of various polymeric based sealant materials.

Areas for Future Research

A review of previous developments made to determine suitable performance related tests for sealants suggests that future research should be centered in two distinct areas:

i) durability of sealants to cyclic movement;
ii) development of analytical methods to evaluate the durability of sealants.

Durability of Sealants to Cyclic Movement

Work undertaken by Karpati has shown that cyclic tests undertaken in the lab can be used to produce deformations in sealants consistent with those obtained from an exposure rack and hence comparable to that known to occur in sealants used in building joints. The results obtained from these tests suggest that provided the type of cycle is properly chosen, cyclic tests can be used to predict the durability of a given type of sealant. Of interest then are the consequences of varying the type of cycle on the deterioration of the sealant bead. Initial results have shown that the rate of cycling can be significantly increased without changing the nature of the results. It would be useful to know the extent to which the rate could be increased before the sealant no longer deteriorates under the same influences as that which causes sealants to degrade in joints. It might also be of interest to determine if further improvements in the test method can be achieved if testing is undertaken under simultaneous temperature control. Finally, tests using the SCR should continue such that the degree to which cyclic movement contributes to the degradation of a sealant in relation to other factors known to cause distress can be estimated for the various types of sealants.

Use of Analytical Techniques

Artificial aging, using various types of weather-o-meter devices, is often used as a method to accelerate the degradative effects brought about by UV radiation, heat and moisture. A sealant has an intrinsic durability to a given degradative action whether it be cyclic movement or the other aforementioned factors. The action of individual effects can be isolated by the proper choice of apparatus and aging conditions and these can be investigated using analytical techniques such as differential scanning calorimetry, dynamic mechanical analysis, thermogravimetric analysis and Fourier transform infra-red analysis. These techniques have been successfully employed in numerous studies related to the degradation of polymeric based materials and more recently have been employed to investigate the aging process in roofing materials [*28*]. Their use in the study of aging of sealants is less well documented. Wolf [*6*] cites a study by Maurer and Brazier (see Ref. *6*) using thermomechanical analysis which indicated that this method can be used to characterize the mechanical properties of sealant samples before and after aging. Such type of analytical techniques can be used as a basis for investigating the synergistic effects of various combinations of aging factors.

Summary

A considerable body of knowledge exists related to assessing the long-term durability of sealants. Initial studies focused on determining whether a specific mechanical parameter could readily be used to predict sealant performance. Although advances were made in this area, the usefulness of this approach was considered limited in consideration of the complexity of the test and attention was then placed on observing the physicomechanical degradation of sealants subjected to natural weathering on an exposure rack. Such type of racks had been used by researchers in the past; however, few efforts were made to establish a link between laboratory testing and cyclic testing, although it was known that such testing was most representative of that which occurs on the outside of buildings. The first investigations established a fundamental aspect related to sealant testing in that it was apparent that the most significant factor in the degradation of sealants was cyclic movement. This provided a means by which cyclic movement was obtained using manually adjusted vises. Results of these trials formed the basis of laboratory fatigue tests that indicated that the results of these types of tests, undertaken in the laboratory under controlled conditions and in a relatively short period of time, could be correlated to those observed to occur on the exposure rack and by extension, simulate that which occurs in a joint on the outside of a building. This was achieved for silicone, polysulfide and polyurethane based sealants of the types currently used to seal joints. Recent advances in assessing the performance of sealants have centered on using these and related cycling techniques in an effort to refine and further improve the test procedure. Future research should endeavor to continue studies related to cyclic movement capability and should also consider the investigation of aspects related to the use of analytical techniques to further enhance the understanding of the interrelationship between factors causing degradation in sealants.

References

[*1*] Karpati, K. K., "Weathering of Silicone Sealant on Strain-cycling Exposure Rack," *Adhesives Age*, Vol. 23, No. 1, 1980, pp. 41–47.

[*2*] Karpati, K. K., "Investigation of Factors Influencing Outdoor Performance of Two-Part Polysulphide Sealants," *Journal of Coatings Technol*, Vol. 56, No. 719, 1984, pp. 57–60.

[*3*] Karpati, K. K., "Performance of Polyurethane Sealants on a Strain-cycling Exposure Rack," *Matériaux et Construction*, Vol. 22, 1989, pp. 60–63.

[*4*] Margeson, J. L., "The Effect of Joint Movement During Cure on Sealant Strength Development," C. J. Parise, Ed. Science and Technology of Building Seals, Sealants, Glazing and Waterproofing, ASTM, STP 1168, American Society for Testing and Materials, Philadelphia, 1992, pp. 22–29.

[*5*] Beech, J. C., Test Methods for the Movement Capability of Building Sealants—The State of the Art, *Matériaux et Construction*, Vol. 18, No. 108, 1983, pp. 473–481.

[*6*] Wolf, A., "Studies of the Ageing Behaviour of Gun-Grade Building Joint Sealants," State of the Art, *Polymer Degradation and Stability*, Vol. 23, 1989, pp. 135–163.

[*7*] Ashton, H. E., "Evaluating the Performance of Organic Coatings and Building Materials," *Permanence of Organic Coatings*, ASTM STP 781, American Society for Testing and Materials, Philadelphia, 1982, pp. 67–85.

[*8*] Karpati, K. K., "A Rational Approach to Building Sealant Testing," *Adhesives Age*, Vol. 16, No. 11, 1973, pp. 27–30.

[*9*] Karpati, K. K. and Gibbons, E. V., "Experimental Prediction of Joint Movement in Buildings," *Materials Research and Standards*, Vol. 10, No. 4, 1970, pp. 16–21.

[*10*] Karpati, K. K., Extension Cycling of Sealants, 12th FATIPEC Congress, Garmisch-Partenkirchen, May 1974, pp. 455–459.

[*11*] Panek, J. R., Cook, J. P., Eds., *Construction Sealants and Adhesives*, 2nd Ed., John Wiley and Sons Inc., N.Y., N.Y. 1970, p. 348.

[*12*] Brown, N. G., Assessment of Joint Sealants by Outdoor Exposure in Cyclic Movement Testers, CSIRO Div. of Building Research, Report No. 01.1-2, Melbourne, 1965, p. 16.

[*13*] Ballantyne, E. R., "Curtain Wall Glazing Sealants," *Build. Decor. Mat.* Vol. 1, No. 5, 1960, p. 40.

[*14*] Ballantyne, E. R. and Brown, N. G., Requirements of Glazing Sealants, Australian Building Research Congress, Melbourne, Paper 1.CS.2, 1961.

[*15*] Fry, J. I. and Whitney, R. S., The Natural Weathering of Sealants in Cyclic Movement Testers, Building Research Association of New Zealand, Judgeford, Technical Paper, 1979, p. 22, p. 26.

[*16*] Sharman, W. R., Fry, J. I. and Whitney, R. S., "Six Years Natural Weathering of Sealants," *Durability of Building Materials* 2, 1983, pp. 79–90.

[*17*] Burstrom, P. G., "Durability and Ageing of Sealants," *Durability of Building Materials and Components*, ASTM STP 691, P. J. Sereda and G. G. Litvan, Eds., American Society for Testing and Materials, Philadelphia, 1980, pp. 643–657.

[*18*] Beech, J. C. and Turner, C. H. C., "Cyclic Joint Movement," *Bldg. Res. & Prac.* Vol. 11, No. 5, 1983, pp. 287–291.

[*19*] Karpati, K. K., Solvason, K. R. and Sereda, P. J., "Weathering Rack for Sealants," *Journal of Coatings Technology*, Vol. 49, No. 626, 1977, pp. 44–47.

[*20*] Karpati, K. K., "Exposure Evaluation of Sealants with Low Movement Capability," *Adhesives Age* Vol. 31, No. 5, 1988, pp. 20–23.

[*21*] Karpati, K. K., "Quick Weathering Test for Screening Silicone Sealants," *Journal of Coatings Technol*, Vol. 56, No. 710, 1984, pp. 29–32.

[*22*] Karpati, K. K., "Device for Weathering Sealants Undergoing Cyclic Movement," *Journal of Coatings Technology*, Vol. 50, No. 641, 1978, pp. 27–30.

[*23*] Karpati, K. K., "Development of Test Procedure for Predicting Performance of Sealants," ACS Symp. Series No. 113, 1979, pp. 157–179.

[*24*] Karpati, K. K., "Testing Polysulphide Sealant Deformation on Vises," *Adhesives Age*, Vol. 28, No. 5, 1985, pp. 18–27.

[*25*] Karpati, K. K., "Laboratory Fatigue Test of a Two-Part Polysulphide Sealant Correlated to Outdoor Performance," *Durability of Building Materials*, Vol. 5, 1981, pp. 35–51.

[*26*] Lacasse, M. A., *Fatigue Testing of Polyurethane Sealants*, to be published.

[27] Matsumoto, Y., Ono, T. and Maruichi, T., "Fatigue and Movement Capabilities of Field-Molded Sealants," *Durability of Building Materials and Components*, ASTM STP 691, P. J. Sereda and G. G. Litvan, Eds., American Society for Testing and Materials, Philadelphia, 1980, pp. 678–686.
[28] Paroli, R. M. and Dutt, O., "Dynamic Mechanical Analysis Studies of Reinforced Polyvinyl Chloride (PVC) Roofing Membranes," *Proceedings of the American Chemical Society, Div. of Polym. Materials: Sci. & Eng.*, Vol. 65, 1991, pp. 362–363.

Nathaniel E. Leonard[1] *and Tariq M. Malik*[1]

The Relation Between Rheology, Artificial Weathering, and Urethane Sealant Performance

REFERENCE: Leonard, N. E. and Malik, T. M., "**The Relation Between Rheology, Artificial Weathering, and Urethane Sealant Performance,**" *Science and Technology of Building Seals, Sealants, Glazing, and Waterproofing: 3rd Volume, ASTM STP 1254*, James C. Myers, Ed., American Society for Testing and Materials, Philadelphia, 1994, pp. 21–32.

ABSTRACT: There has been vigorous debate on the ability of urethane sealants to withstand the conditions found in high-performance construction applications. A number of studies have suggested that multicomponent urethanes, as a generic technology, are unable to endure the high heat, humidity and ultraviolet light found in some parts of the world.

This paper argues that it is incorrect to make blanket statements about the performance characteristics of *any* sealant technology, whether it is silicone or organic. The focus is, however, on multicomponent urethanes. Results of standard ASTM xenon and QUV tests are discussed. Several novel test regimens, including the "frog pond test" and sophisticated rheological techniques, are also reviewed. To gather as much data as possible, samples were run through these procedures for as long as 7000 hours.

This paper outlines several important conclusions. First, silicones actually have a *higher* modulus at cold temperatures (below −35°C) than the multicomponent urethanes examined. Second, the wide range of performance results in the urethanes tested highlights the fallacy of blanket statements about these sealants. Some were clearly unable to endure the conditions to which they were exposed. Other multicomponents, however, did not degrade. Rheological techniques are used to augment the data provided by the traditional QUV and xenon procedures.

KEYWORDS: rheology, frog pond, multicomponent urethanes, silicones, artificial weathering, QUV, xenon, carbon arc

To qualify as a high-performance construction sealant, a material must meet the requirements of ASTM C 920 Specification for Elastomeric Joint Sealants. In general, this means the product must deliver the following key features: resistance to ultraviolet light; low temperature flexibility; user-friendly mixing (for multicomponents), gunning and tooling; minimal sag at high temperatures; quick tack-free time to reduce dirt pick-up, long-term movement and adhesion capabilities; high solid content to ensure low shrinkage; and acceptable shelf life.

In recent years there has been concern whether *any* multicomponent urethanes could meet these imposing requirements [*1*]. This paper presents a variety of analytical techniques that examine this charge's validity. As a starting point, results from and the relevance of standard ASTM tests are reviewed. Shortcomings in these procedures are discussed and more sophisticated techniques are then described. Finally, and *most* importantly, the unique insights and data offered by rheology are outlined. The key finding of this research is that all multicomponent

[1]Mr. Leonard is Market Manager and Dr. Malik is Research and Development Associate/Group Leader, Tremco, Inc., Sealants and Coatings, 3735 Green Road, Beachwood, OH 44122.

urethanes are *not* alike. Some are clearly inadequate. Others deliver the performance required given today's construction techniques.

QUV Weatherometer

Three cement-color commercially available Class A multicomponent urethane sealants and one black commercially available one-component silicone were included in this portion of the study. Five patties on unprimed aluminum panels were made of each sealant, with each sample approximately 2.5 × 70 × 90 mm.

The specimens were cured for 72 hours at 25°C, 50% relative humidity. Four were placed in a QUV weatherometer equipped with UVB-313 lamps. The fifth specimen was kept at ambient conditions as a control. Light and condensation cycles were run according to ASTM G 53 Practice for Operating Light- and Water- Exposure Apparatus (Fluorescent UV-Condensation Type) for Exposure of Nonmetallic Materials. The temperature during the procedures was 60°C, ±5°C. Microscopic evaluations were conducted at 25 times after 600, 1000 and 3500 hours exposure in the weatherometer. ASTM C 920 only requires 250 hours, an extremely short exposure level.

As shown in Figs. 1 to 3, the products showed a wide endurance range. Sealants A (polyurethane A, or PLA-A for short on the Figures), B (PLA-B) and C (PLA-C) are two-part chemical cure polyurethanes and D is a neutral-cure, medium-modulus silicone. A and B exhibit cracking,

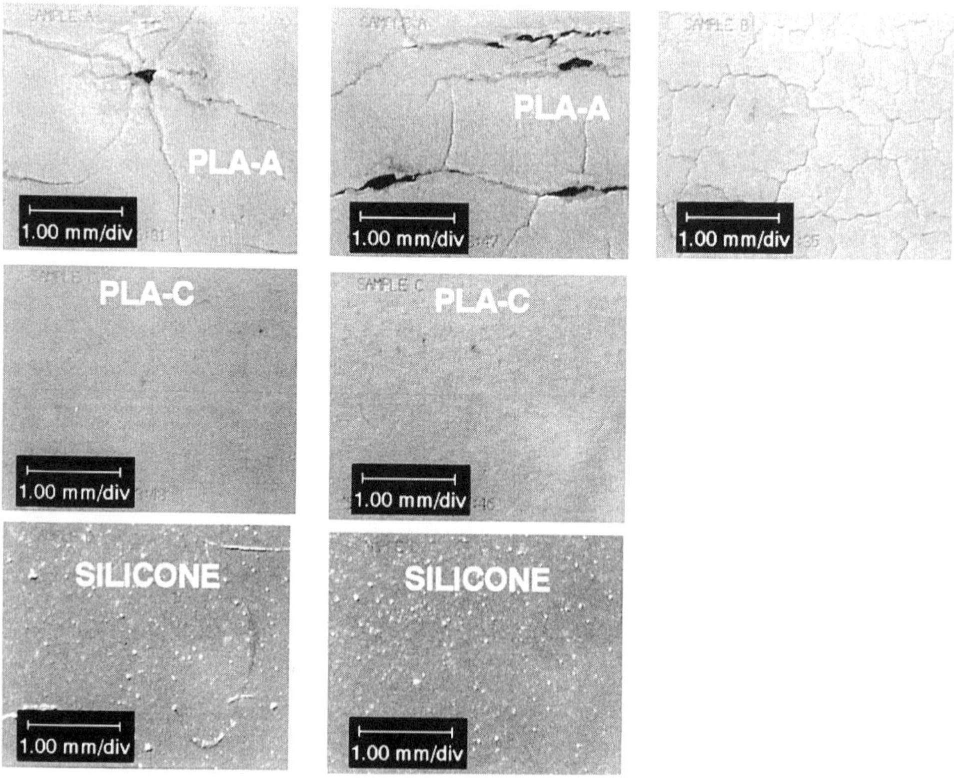

FIG. 1—*Effect of accelerated weathering (600 hours, Q-U-V) on polyurethane and silicone sealants.*

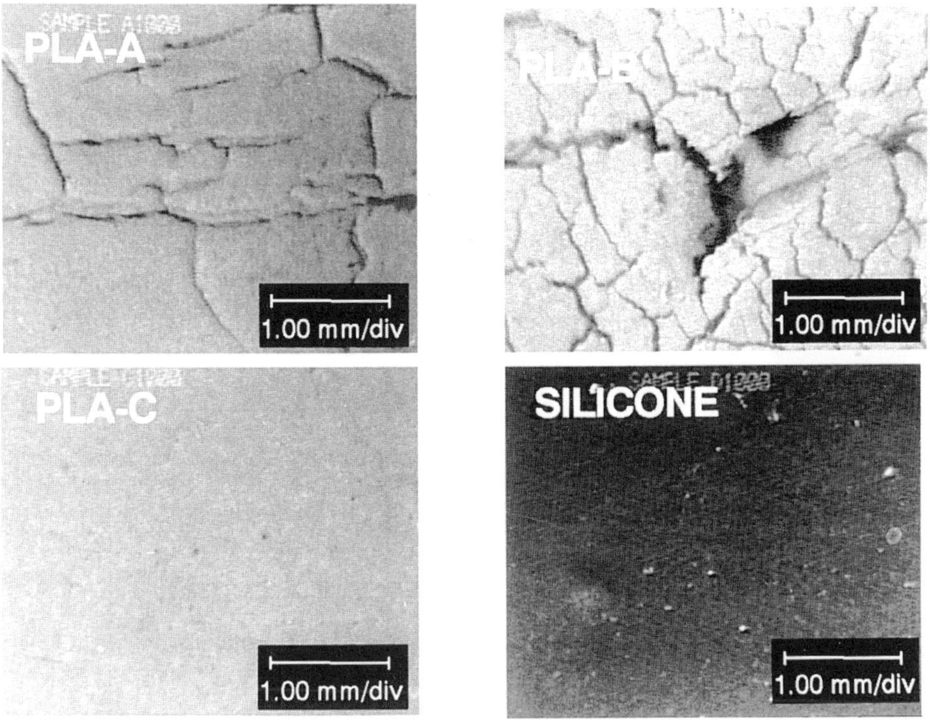

FIG. 2—*Effect of accelerated weathering (1000 hours, Q-U-V) on polyurethane and silicone sealants.*

FIG. 3—*Effects of accelerating weathering (3500 hours, Q-U-V) on polyurethane and silicone sealants.*

blistering and bubbling that become more severe as the QUV exposure is increased. In contrast, neither C (a urethane) nor D (a silicone) show any significant deterioration.

(Note: Additional multicomponent urethanes, E-H, are used in later sections of this paper. The initials are consistent throughout the research, that is, the same product is always A, etc. Due to resource limitations, it was impossible to run all the products through the entire range of procedures described in this document.)

Some initial work has been done to correlate the QUV results and the other data presented here with performance on actual buildings. These preliminary examinations support the implication of the QUV data that C and D are more durable than A and B. Several jobs in high heat, high humidity environments have been tracked where either A or B were applied and failed. In the same cities, sites with C or D have performed well. In a few cases, A and/or B deteriorated and were successfully replaced by C. (The buildings involved were in Phoenix, Arizona and had red granite panels and sealant. Reversion occurred after three years and was apparent on the south and west elevations.) More systematic work is required, however, to confirm a correlation.

In spite of QUV's usefulness, this analysis has several limitations. First, in spite of the field analysis just described, the exact correlation between QUV results and actual job performance varies by product due to the different way each compound reacts to the artificial light wavelengths. Second, QUV does not measure adhesion, movement or handling characteristics.

Xenon Weatherometer

To validate these QUV results, xenon weathering tests were done according to ASTM G 26 Practice for Operating Light-Exposure Apparatus (Xenon-Arc Type) With and Without Water for Exposure of NonMetallic Materials. Samples of polyurethane products A, C and E were weathered for 7000 hours and removed so photographs could be taken. As Fig. 4 shows, there is a significant difference between the three samples. Product A has developed deep mudcracks but urethanes C and E have suffered only minimal damage. Sample C, in fact, only has fine, superficial unidirectional cracks. These results follow the same pattern as

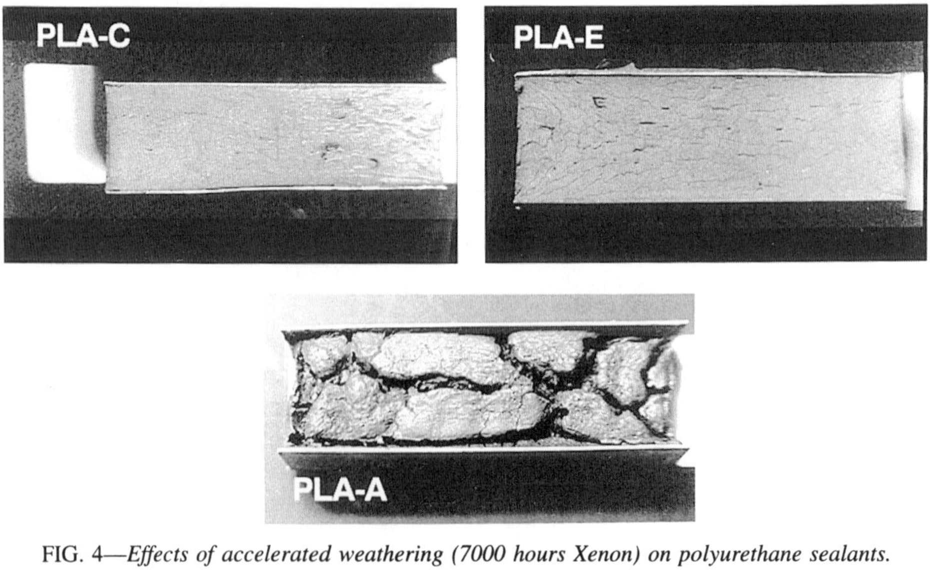

FIG. 4—*Effects of accelerated weathering (7000 hours Xenon) on polyurethane sealants.*

those of the QUV study. This highlights the importance of differentiating between materials, regardless of the similarity of their generic technology class.

Extension Analysis

Although interesting, the xenon study suffers some of the same weaknesses as the QUV approach. Most importantly, adhesion and movement properties are not examined in standard xenon and QUV procedures. Previous papers have focused on this issue and we decided to adopt one common extension evaluation technique here [*1*].

Six 12 mm × 12 mm × 50 mm samples each of polyurethanes C and F without backer rod were prepared using the H-bead configuration without primer. After being given four weeks to cure at 25°C, 50% RH, each sample was extended 25% of initial joint width in an Instron tensile tester, blocked then placed in an Atlas CXW Weather-ometer. The machine was set on a repeating cycle of 108 minutes with light only and 12 minutes with both light and water spray. The light's intensity was 0.35 W/m^2 and temperature was 70°C, ±5°C.

The samples were weathered 930 hours and then examined. Surface degradation was minimal on all the samples, although partial loss of bond was observed in specimens of both C and F sealants.

These extension results were confirmed by a separate set of studies that subjected three unprimed product C and three unprimed product F extended samples to 1500 hours in a carbon arc weatherometer followed by 9000 hours in the xenon weatherometer. As shown in Fig. 5, several samples did suffer adhesive failure. This is not totally unexpected because of the length and severity of the exposure and the fact that none of the samples were primed. It is important to note, however, that *none* of the samples underwent catastrophic reversion or softening. C did stiffen, but as the rheological section will discuss, not to problematic levels. Again, this information strengthens the hypothesis that all sealants are *not* alike.

Unfortunately, this test method, although intriguing, is only relevant to actual job-site conditions in the rare situation where a building is caulked on the single hottest day of the year. Continued high UV exposure is typically found in hot environments, such as Florida, Texas, or Arizona. High heat causes substrates to expand and *closes* the sealant pocket. The 25% extension/high UV combination, therefore, is analyzing a property that does not deliver significant value to architects and building owners. This procedure has been included in this paper because it is being considered by several industry associations, ISO in particular, for inclusion in specifications.

Frog Pond Analysis

None of the procedures described previously directly examine the conditions that contribute to reversion, or the depolymerization of sealants. Tremco believes that the factors that lead

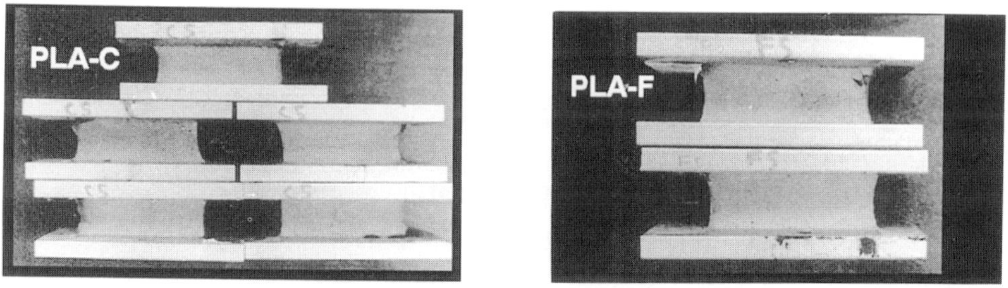

FIG. 5—*Sample extension analysis 25% extension after 1500 hours carbon arc, 9000 hours Xenon.*

to this phenomenon are, in order of importance, heat, humidity and ultraviolet exposure. To test sealants' ability to withstand these harsh conditions, the "frog pond" test was developed.

The frog pond, per Fig. 6, aims S-1 sunlamp bulbs at 50-mm thick sealant patties with backer rod placed between the material and a piece of wood, which is placed in a thin layer of water. The distance between the lamp was approximately three inches, but varied as water evaporated. Temperatures varied between 70° and 90°C, the samples were water-saturated by the backer rod and the wood. Humidity was uncontrolled because the apparatus was not completely enclosed. These conditions are meant to be similar to those found in a non-vented cavity wall in hot climates (such as Texas or Florida) where water can collect behind the sealant, especially if an open-cell backer rod is employed. Unlike the QUV, samples in the frog pond are exposed to water saturation.

Two concrete-color samples of polyurethanes A,B, C and H were put in the frog pond for 90 days. As Fig. 7 shows, there is a wide range of visual results. Sealants A, B and H are clearly depolymerizing or reverting. C, however, is not demonstrating any visible change and closely resembles D (the silicone). These results are consistent with the data collected from the procedures already discussed. A performed poorly in the QUV and xenon tests and B also suffered deep cracks in the QUV experiments. C, on the other hand, suffered minimal damage in all the previous research methods.

Rheological Analysis

Polymer sealants exhibit viscoelastic behavior, which means that these materials have some of the characteristics of both viscous liquids and elastic solids. Elastic materials have the capacity to store mechanical energy with minimal dissipation. Viscous liquids, on the other hand, dissipate, and are unable to store, energy.

When sealants are deformed during joint movement a portion of the applied energy is stored as potential energy and a portion is dissipated as heat. This simultaneous viscous and elastic behavior is dependent on stress, strain, time (or frequency), temperature and environmental

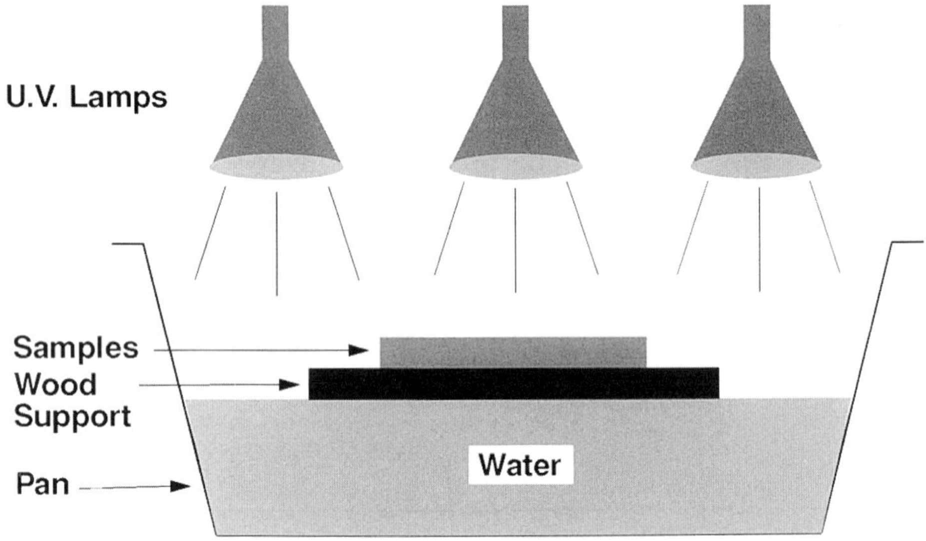

FIG. 6—*Frog pond.*

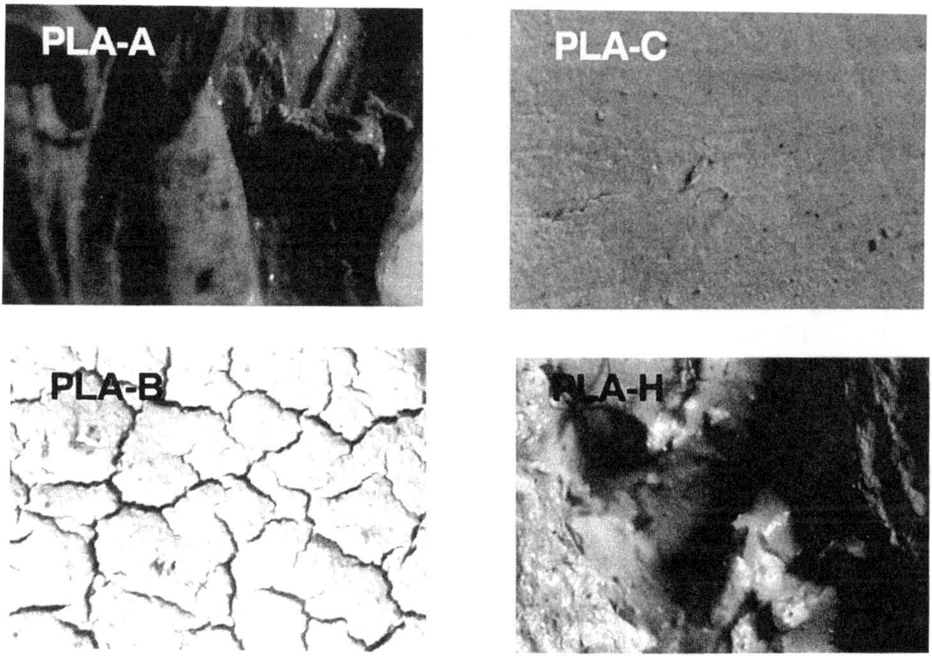

FIG. 7—*Frog pond samples after 90 days.*

conditions. Rheological measurements, and dynamic mechanical analysis in particular, provide valuable quantitative information on the viscoelastic properties of sealants and the effect of the parameters mentioned above on long-term performance.

Measurements are usually taken by applying a sinusoidal load to the sample and calculating a real or in-phase modulus (G'), and an imaginary or out-of-phase modulus (G''). The real and imaginary moduli represent the elastic and viscous behavior of the sample and are called, respectively, storage and loss moduli. The ratio of loss modulus to storage modulus (G''/G') defines the loss tangent factor, often referred to as tan delta [*2,3*]. The ability to measure both of these moduli permits a much fuller characterization of sealants. Other than perfect elastomers (sealants are not perfect), if a material is subjected to a constant load, the elastic modulus exhibited will decrease over a period of time. This occurs because polymeric materials under a load will undergo molecular rearrangement in an attempt to minimize the localized stresses [*4*]. Modulus measurements performed over short times result in high values, whereas measurements taken over long times result in low values.

The effect of time can also be observed by the change of frequency of deformation under which sealants are subjected. At a given temperature, a sealant deformed at high frequency will have a relatively high modulus whereas deformation at low frequency will result in low modulus.

Similarly, temperature plays a crucial role in sealant performance. Sealants are soft at room temperature but exhibit a large increase in stiffness as they are cooled below their glass transition temperature. The effects of time and temperature must be taken into account in the selection of sealants for structural and construction applications. Dynamic mechanical analysis can predict the effects of time and temperature on sealant performance under different environments.

Sample Preparation and Test Procedure

Products A, B, C, and D were included in the study. Each sample was prepared according to ASTM C 793 Test Method for Effects of Accelerated Weathering on Elastomeric Joint Sealants and was 2.5 × 70 × 90 mm. Five samples were made of each product and cured for 72 hours. Conditions were 25 ±5°C, 50 ± 4% RH.

One sample of each product was maintained at standard conditions as a control and the other four were put in a QUV weatherometer with UVB-313 lamps and cycled per ASTM G 53. Rheological readings were taken at several points during the experiments, and the specimens were left in the QUV for a total of 3500 hours.

A Rheometrics Mechanical Spectrometer (RMS 605) was used for cold-temperature experiments and a Bohlin Constant Stress rheometer was used to generate all the other rheological data. All measurements were made at 1 rad/s and 0.5% strain level. A constant heating and cooling rate of 1°C/min was used and measurements were taken at each 3°C interval, with 1 min soak time.

Acceptable Performance Range

High-performance construction sealants should maintain their dynamic storage modulus (G′) across a wide temperature range. Figure 8 shows the acceptable parameters for most common control and expansion joint applications (based on Tremco, Inc. research). At low temperatures, the storage modulus should not exceed 10^8 dyn/cm^2 because higher levels would make the sealant too stiff to withstand significant movement for extended time periods. (In Canada and the northern United States, air temperature can reach −35°C in Winter. Surface temperature in Winter is typically 5°C lower than ambient levels [5]). At room temperature,

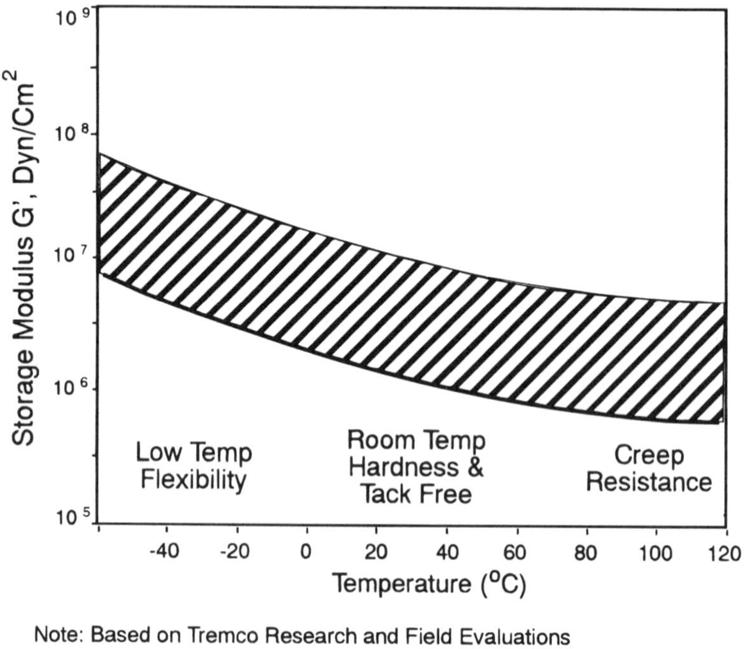

FIG. 8—*Acceptable sealant modulus range.*

sealants should be 5×10^6 to 5×10^7 dyn/cm^2. Higher levels lead to poor adhesion and lower values increase the risk of unacceptably long tack-free times. At high temperatures, the storage should be between 10^6 to 10^7 to ensure good creep and sag resistance.

Cold Temperature Performance

Figure 9 indicates that silicones are stiffer than urethanes at low temperatures below −40°C. Silicone's sharp modulus increase at −40°C is probably caused by the crystalline melting of polydimethylisiloxane polymer. In general, therefore, *all* silicones with this polymer will show this fundamental shift. Other researchers have come to the same conclusion [*6–8*]. This jump is *not* caused by the glass transition, which occurs at about minus 100°C.

Weathered Sample Results

The rheological data shown in Fig. 10 confirms the photographic QUV results which showed severe cracking of samples A and B and the excellent durability of sealants C and D. In this test, temperature was scanned from 25°C to 125°C, held for two hours then returned to 25°C. An ideal sealant would have a storage modulus independent of temperature and thermal history.

Urethanes A and B do not meet this requirement. In both cases their storage moduli at high temperatures have fallen to dangerously low levels. In addition, at cold temperatures they have become unacceptably stiff. Samples C and D, however, are stable across the entire range. This powerful *rheological* technique provides quantitative evidence that, as indicated by the xenon and QUV tests, there is a significant difference between urethanes A, B and C.

The 1000 and 3500 hour results support these conclusions. (See Figures 2, 3 and 11, 12.) A and B continue to deteriorate, clearly losing their ability to withstand movement at high temperatures. The flat curves for C and D, on the other hand, indicate that these products have undergone no rheological degradation. The thermal and UV stresses from the QUV procedures have had minimal impact on their storage modulus.

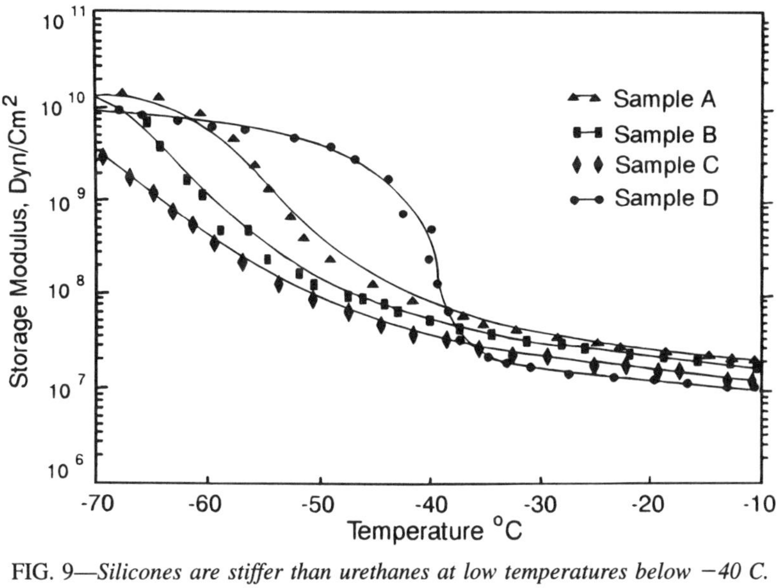

FIG. 9—*Silicones are stiffer than urethanes at low temperatures below −40 C.*

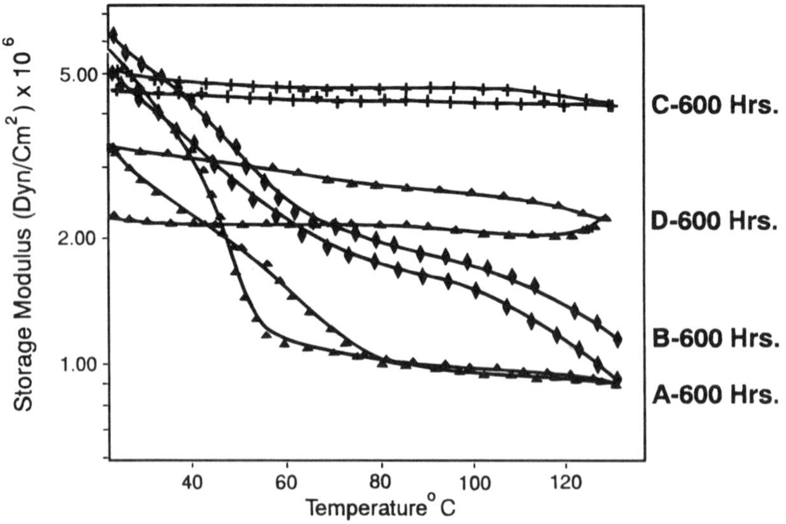

FIG. 10—*Rheological data.*

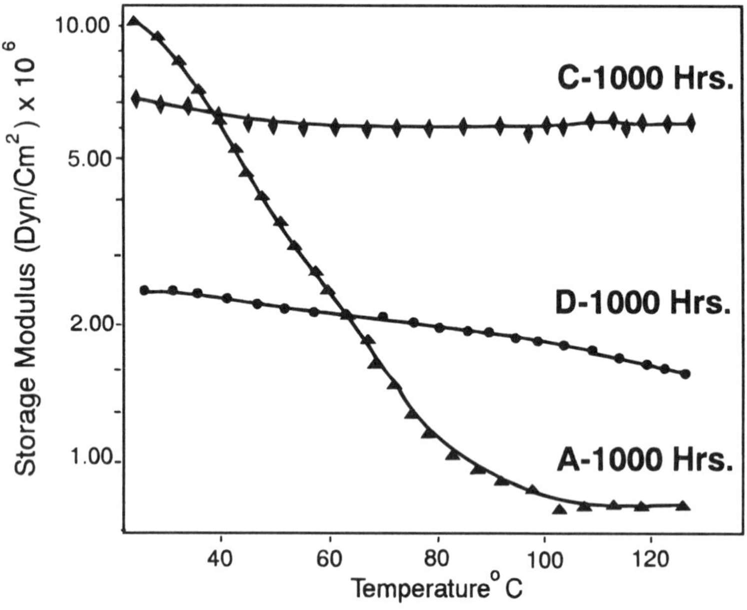

FIG. 11—*1000 hour results.*

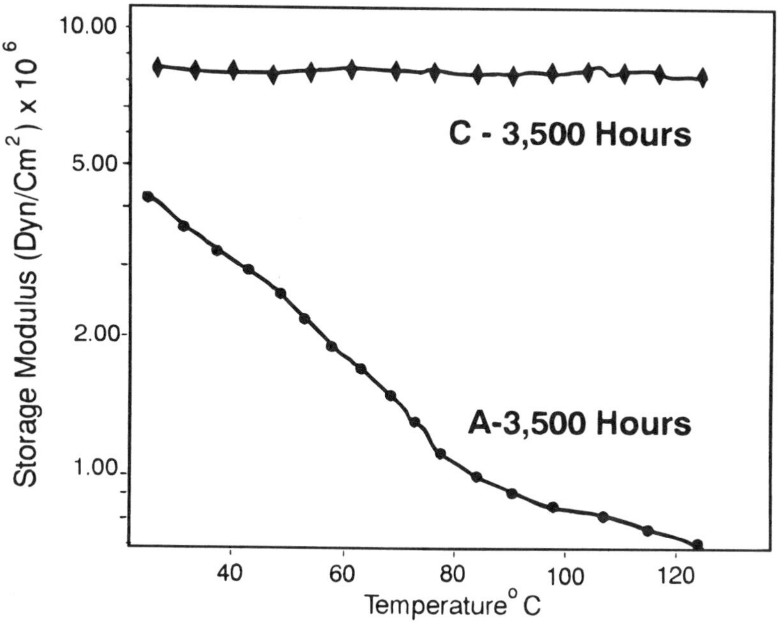

FIG. 12—*3500 hour results.*

Conclusions

- Based on limited testing, at least one silicone sealant is much stiffer than the polyurethanes examined in this paper at temperatures below −35°C(−31°F). More work is required to understand this issue.
- Some commercial polyurethane sealants show decomposition, softness and reversion at temperatures above 80°C and stiffness at temperatures below −40°C after QUV exposure.
- One of the four multi-component urethanes tested did not revert and shows similar performance to silicone sealants under severe, quantifiable, test conditions.
- No sealant technology—whether it be polyurethane, silicone or another chemistry—should be characterized as a generic class.
- The current tests incorporated in ASTM C 920 Specification for Elastomeric Joint Sealants do not identify the critical performance differences between sealants and must be updated to accomplish this critical task.

Next Steps

- Analyze a wide range of silicone sealants to evaluate differences in their performance characteristics.
- Continue development of a number of laboratory experiments that more closely duplicate actual field conditions.
- Continue field work to correlate laboratory and actual job data.
- More specific research on the causes of reversion.

Acknowledgments

We would like to thank Steven L. Hannah, Group Manager, Tremco Research and Development, Sealants and Coating, for organizing most of the QUV, xenon and carbon arc experiments.

References

[*1*] Bridgewater, T. J. and Carbary, L. D., "Accelerated Weathering and Perimeter Sealants," *Science and Technology of Building Seals, Sealants, Glazing and Waterproofing: Second Volume, ASTM STP 1200*, Jerome M. Klosowski, Ed., American Society for Testing and Materials, Philadelphia, 1992.
[*2*] Duncan, J. C., ANTEC, 1989, p. 543.
[*3*] Ferry, J. D., *Viscoelastic Properties of Polymers*, Third Edition, Wiley, New York, 1980.
[*4*] Sichina, W. J. and Gill, P. S., ANTEC, 1989, p. 1111.
[*5*] Stephenson, D. G., "Canadian Building Digest," No. 47, Division of Building Research, National Research Council of Canada, November, 1963.
[*6*] Hourston, D. J. and Zrandous, M., *Advances in Interpenetrating Polymer Networks*, D. Kempner and K. C. Frisch, Eds., Technomic Publishing Co., 1990, p. 191.
[*7*] Novikov, A. S., *Soviet Rubber Technology*, 12, 1960, p. 4.
[*8*] Helmer, J. D. and Palamteer, K. E., *Journal of Applied Polymer Science*, Vol. 13, 1969, p. 2112.

John Beech[1] *and John Beasley*[1]

Further Studies of Cure and Durability of Building Sealants

REFERENCE: Beech J. and Beasley, J., "**Further Studies of Cure and Durability of Building Sealants,**" *Science and Technology of Building Seals, Sealants, Glazing, and Waterproofing: 3rd Volume, ASTM STP 1254*, James C. Myers, Ed., American Society for Testing and Materials, Philadelphia, 1994, pp. 33–50.

ABSTRACT: Some results from the second phase of a two year experimental study are reported and discussed. A range of well cured commercial sealants products was exposed to natural weather and to various laboratory simulations of the factors in weather which cause degradation or changes in properties, for up to 12 months. These include heat (70°C), with and without UV(B) radiation, and artificial weathering including UV(B) radiation and moisture condensation (QUV apparatus).

The cure properties of polyurethane, silicone and polysulphide sealants over a two year period are described. For these materials, the majority of which were well cured in the second year of the study, the effects of heat and of UV(B) radiation are discussed. Polyurethane sealants showed significant correlations between changes in modulus and in extensibility, after exposure to UV and to artificial weather. The effect of heat was more significant for polysulphides and for some silicone sealants.

As in the earlier phase of this work it was possible to demonstrate significant correlations between modulus changes in the full range of sealants after exposure to weather for twelve months and to artificial weathering for eight weeks. In addition, heat aging caused changes in extensibility that correlated highly with the effects of natural weather.

KEYWORDS: sealants, cure, durability, weather, artificial weather, property changes

A long term objective of research into the performance of sealants at the Building Research Establishment (BRE) has been the development of methods of test for the durability of these materials in service. The work has aimed to simulate in the laboratory the conditions of service of a building joint, involving the individual factors believed to cause degradation of sealants, and appropriate combinations of these.

Early attempts to correlate changes in key properties after these laboratory exposures, with those of sealants specimens exposed to natural weather, were frustrated by the effects of the slow curing of many of the sealants products employed in the tests. It was apparent that some of these were still undergoing marked changes in modulus twelve months after preparation of the samples, so that it was difficult to establish a datum condition with which material exposed to natural or to artificial weathering could be compared.

Consequently, in 1990 a test program in which a large number of specimens of a range of 9 commercial sealants were prepared, was started. Specimens were to be tested according to a strict timetable after exposure to weather at Garston, England and to various laboratory test regimes.

[1]Building Research Establishment, Garston, Watford, Hertfordshire, WD2 7JR, England.

The work was divided into two phases. In the first freshly prepared sealants specimens were employed; while in phase II the test procedure was repeated using a second set of specimens that had been stored for 12 months under standard laboratory conditions. By this means it was intended to distinguish more clearly between property changes due to the continuing cure processes of the various sealants, and those changes caused by exposure to the various degradative factors of the service environment.

The results of Phase I of the experimental program were reported in part in a paper [*1*] presented at the ASTM Second Symposium on Sealants in 1992 (Fort Lauderdale). Large differences in the respective rates of cure of control specimens of the various sealants tested were found as indicated by measurements of the tensile moduli of the materials. It was evident that this must be allowed for in any test procedures designed to assess the long term performance of building sealants. It was shown that some highly significant correlations existed between modulus values measured on artificially weathered samples and those of sealants exposed to temperate weather for 12 months.

This paper reports the results of the second phase of the study, in which well cured specimens were tested. As in previous work at BRE to assess the long term performance of building and construction sealants, we have selected two parameters derived from tensile test data as 'performance indices': the secant modulus (as the force at 25% extension) and the extensibility (taken as the percentage elongation at maximum force).

It is proposed that any evaluation of sealant performance should be based on at least two indicative property changes of this kind.

Experimental

Materials

Nine commercial sealants were used in the tests. The coding below is used in the paper.

P13	Black	Moisture curing one part polysulphide to BS5215:1983 [*2*].
2P10	Grey	Two part chemically cured polysulphides to
2P15	Black	BS4254:1986 [*3*].
S6	Grey	Low modulus silicone sealants to
S12	Brown	BS5889:1989 [*4*].
S13	Black	High modulus silicone sealant to BS5889:1989.
U1	Grey	One part polyurethane sealants.
U6	White	
U3	Grey	Multi part epoxy polyurethane.

Preparation of Specimens

The test specimens each comprised a 12 × 12 × 50 mm bead of sealant between the large surfaces of cement mortar and aluminium adherents. The cement mortar pieces were 12 × 50 × 8 mm in dimension and were prepared by the method described in BS4254. Aluminium test pieces were 12 × 50 × 10 mm, with a milled finish.

Five replicate specimens of each sealant for each of the test exposures, were prepared using the manufacturers' recommended primers for the two substrates. The specimens were stored at 23°C and 50% rh (at 27°C and 95% rh in the case of P13) to allow initial cure to take place, drawing the spacer bars back from the sealant after a few days in the case of the moisture curing materials.

Test Exposures

The specimens were allocated randomly to the exposures listed as follows, for each set of specimens made using one of the nine products. As mentioned previously a second set for each product was set aside for 12 months storage at standard conditions, for use in the second phase of the study.

The test exposures used were:

Control (CN)—stored at standard conditions of 23°C ±2°C and 50 ±5% rh.

Heat (HT)—storage in an oven at 70°C ±2°C.

Ultra Violet Light Only (UO)—continuous exposure to UV(B) radiation in a QUV apparatus, with temperature maintained at 70°C.

Ultra Violet Light with Condensation (UC)—artificial weathering in a QUV apparatus with cycles of 4 hours of UV(B) radiation at 70°C followed by 4 hours condensation at 40°C.

Natural Weather (NW)—samples were exposed facing south on racks inclined at 45°, on the weathering site at Garston (a temperate climate in southern England).

This coding is used throughout the paper.

Tensile Testing

Five specimens of each sealant from each of the five test environments were removed after the following durations of exposure: 2, 4, 8, 16, 28, 40 and 52 weeks. They were conditioned for 24 hours at standard conditions before being placed in a tensile test machine and pulled to destruction at 5 mm/min. Force data were recorded at 0.1 mm increments of strain and stored on disc for subsequent analysis.

A set of five control samples of each sealant was also tested in this manner after the initial storage period of 28 days under standard conditions.

From the tensile test data the following parameters were obtained for each specimen: force (N) at 15 mm joint width, which was taken as equivalent to the secant modulus at 25% extension; the maximum force (N) on the force/extension curve; and the extension at which maximum force was attained, and hence the percentage, elongation at this force.

Results and Discussion

Cure Phenomena

With the completion of Phase II of the study, data are available for the change of modulus of the control specimens, during the period from 28 days after preparing them through two years storage in standard conditions. Experimental data for the changes during the first year

(Phase I) are given in reference [1]. Table 1 gives the mean force at 25% extension: initially and after storage for 52 and 104 weeks.

A number of sealants showed appreciable increase in modulus during the first 12 months, notably the one component polysulphide (P13) and one of the low modulus silicones [12]. However, it is clear that virtually all the sealants (except P13) were fully cured after two years, since the differences in mean modulus between the 52 and 104 weeks specimens were not statistically significant. Rates of cure (Table 1) during the second year indicate that 'plateau' modulus values were achieved during this time.

The influence of cure processes on the observed changes in properties measured during this second phase of the study is therefore, of markedly less importance than for the earlier results, reported previously. This is shown most strikingly in Figs. 1 to 4, for the one component polysulphide sealant (P13) for which changes in both extensibility and modulus exhibit different patterns in the first and second year data.

Durability Effects

Graphical data illustrating the changes in the two performance indices, modulus (F25) and extensibility (as the width at maximum force), are given in Figs. 1 to 14. These relate mainly to the data for year II of the study. The equivalent for year I, for all the sealants tested, are shown in reference [1].

Some general observations can be made regarding the nature and magnitude of the changes. The changes in both extensibility and modulus are smaller with exposure to weather than those in the first phase, since they are due almost entirely to the effects of continuing cure. Differences in measured mean values between the naturally weathered samples and same age controls are generally small in magnitude, and only achieve significance (usually at the 5% level) after longer periods of exposure.

TABLE 1—*Modulus changes in control specimens.*

	Mean Force (N) at 25 per cent extension								
	Polysulphide Sealants			Silicone Sealants			Polyurethane Sealants		
Time	P13	2P10	2P15	S6	S12	S13	U1	U3	U6
Initial	8.3	60.4	57.6	27.4	25.2	143.2	54.4	36.7	36.5
At 52 weeks (phase I)	124.2	73.0	81.8	79.3	105.0	161.5	130.2	67.4	72.0
At 104 weeks (phase II)	143.6	69.5	83.0	84.5	94.2	171.6	141.9	67.9	70.3
Rates of Cure[a] in Phase I	0.81	0.18	0.29	0.61	0.85	0.11	0.53	0.45	0.50
Phase II	0.14	−0.05	0.01	0.06	−0.11	0.06	0.08	0.01	−0.02
% of final modulus values: Initial	6%	83%	70%	35%	24%	89%	42%	54%	51%
At 52 weeks	86%	(100%)	99%	94%	(100%)	94%	92%	99%	(100%)

[a] Expressed as the change in mean force at 25% extension, during each 52 week period, as a percentage of the final "plateau" value.

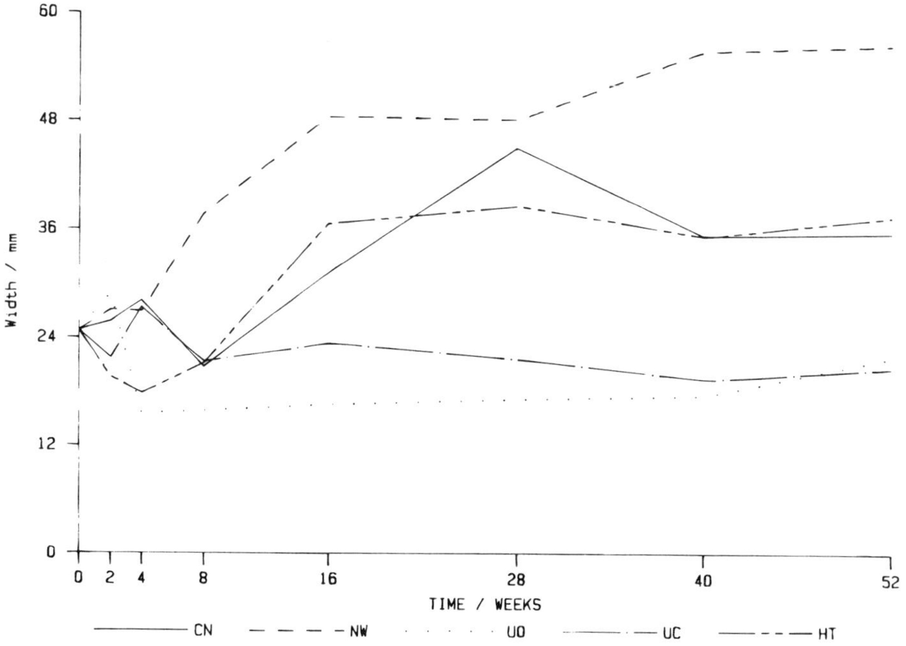

FIG. 1—*P13 Year 1 width at maximum force.*

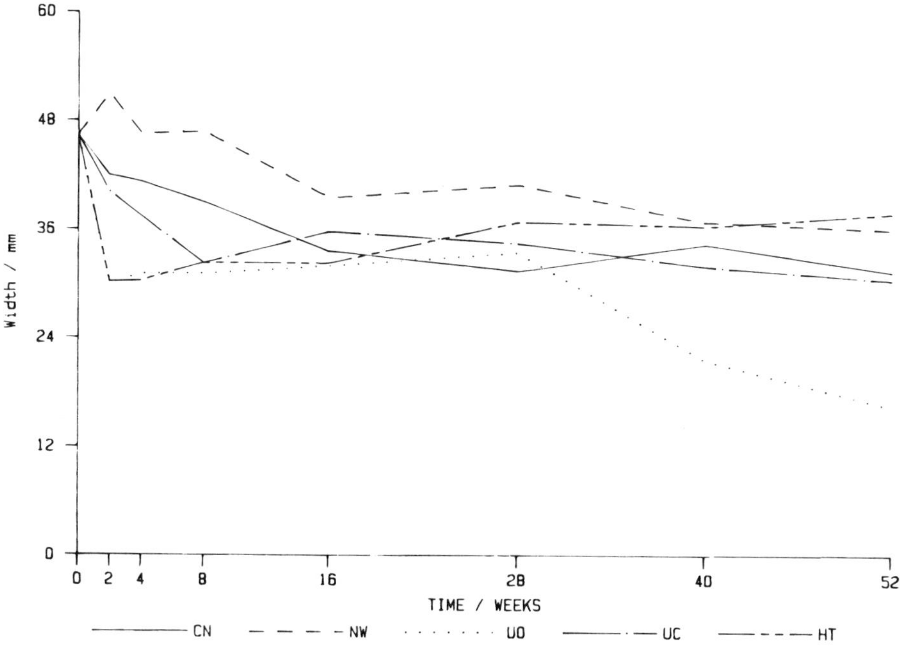

FIG. 2—*P13 Year 2 width at maximum force.*

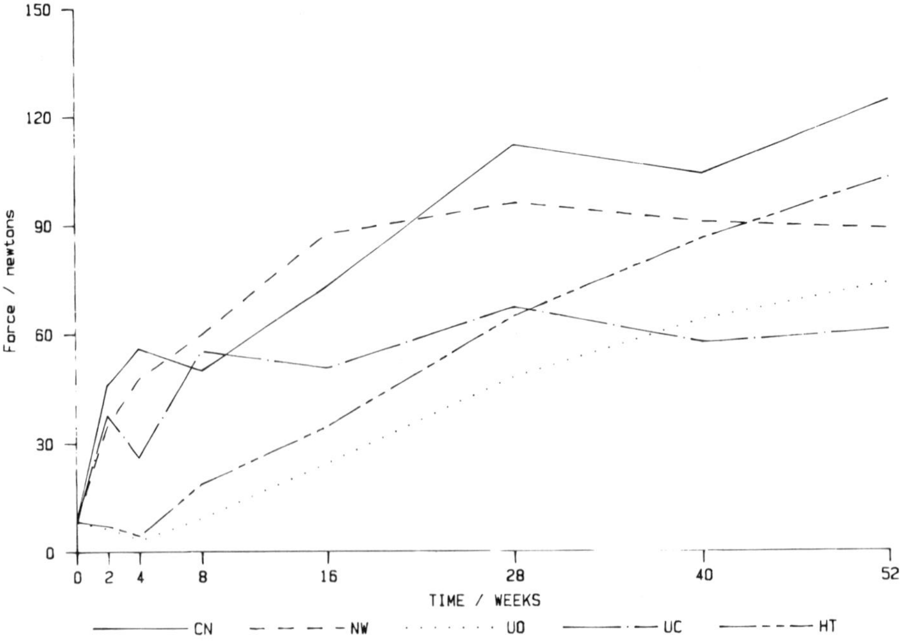

FIG. 3—*P13 Year 1 force at 15 mm width.*

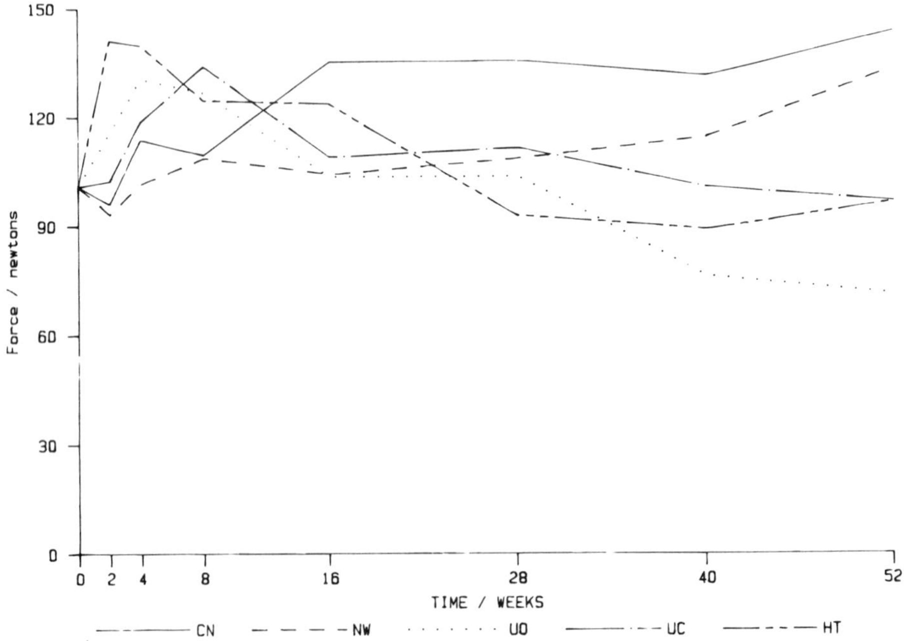

FIG. 4—*P13 Year 2 force at 15 mm width.*

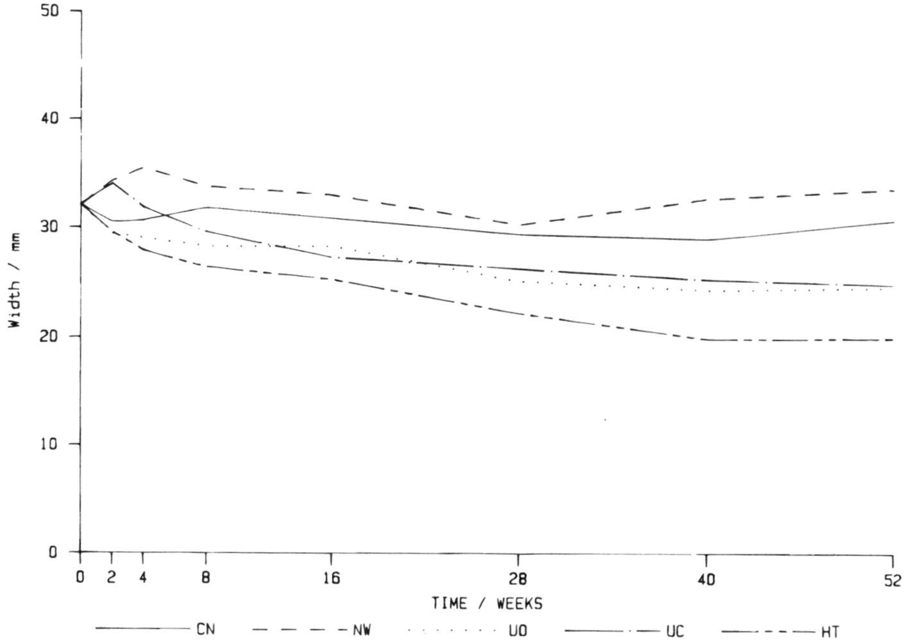

FIG. 5—*2P15 Year 2 width at maximum force.*

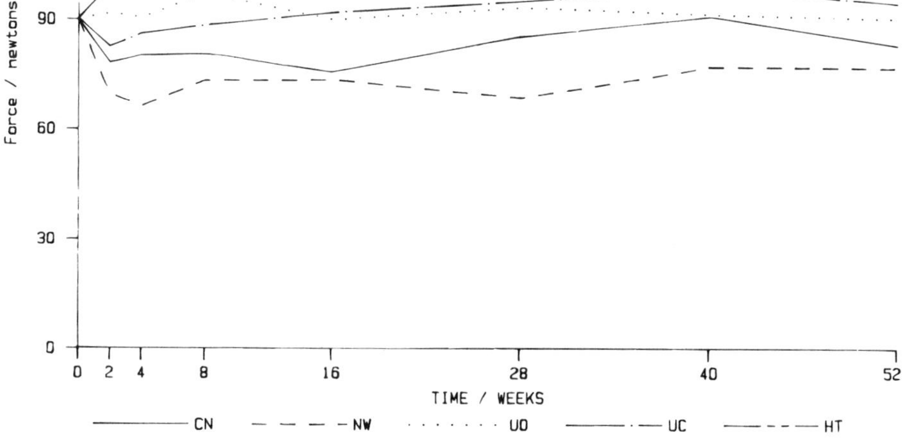

FIG. 6—*2P15 Year 2 force at 15 mm width.*

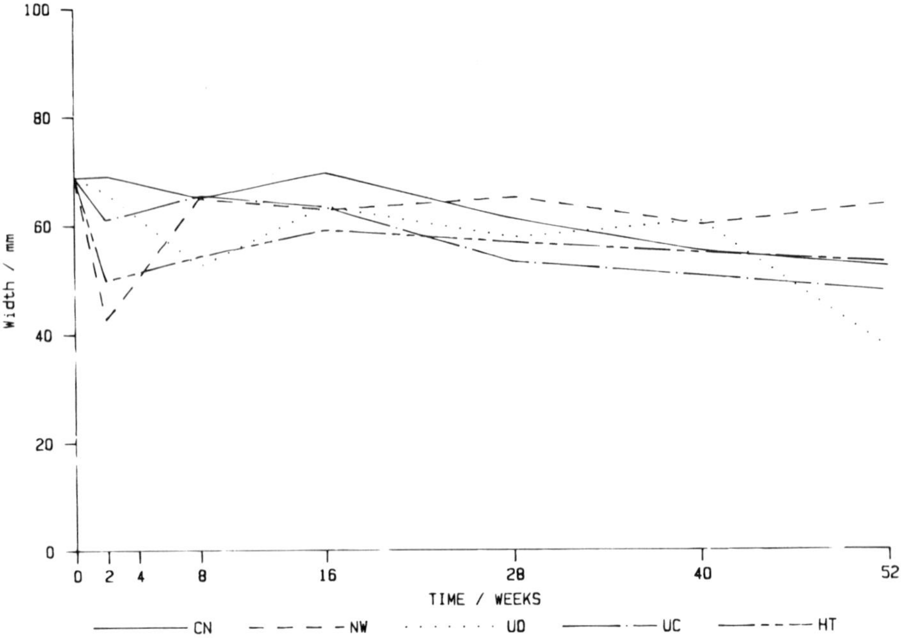

FIG. 7—*S12 Year 2 width at maximum force.*

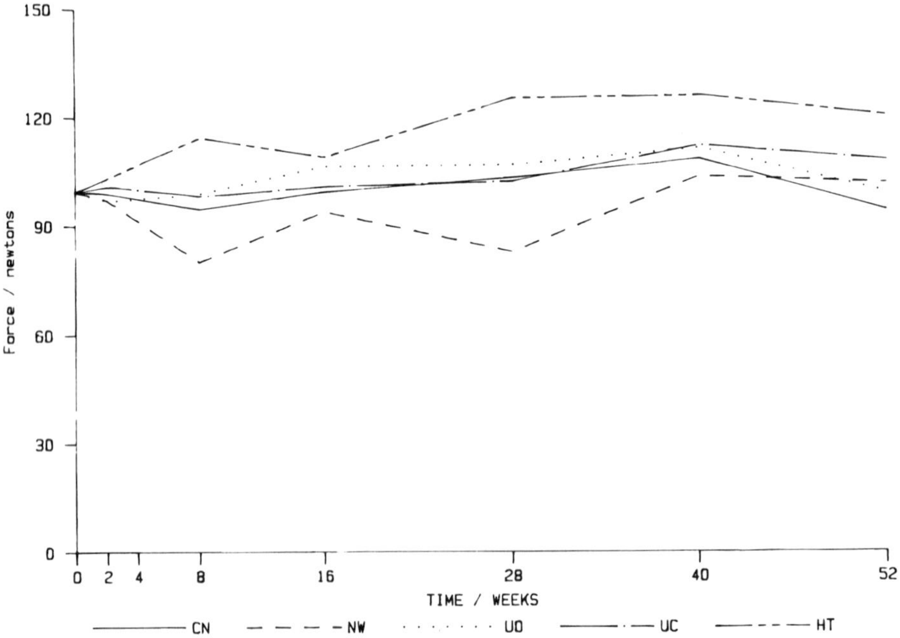

FIG. 8—*S12 Year 2 force at 15 mm width.*

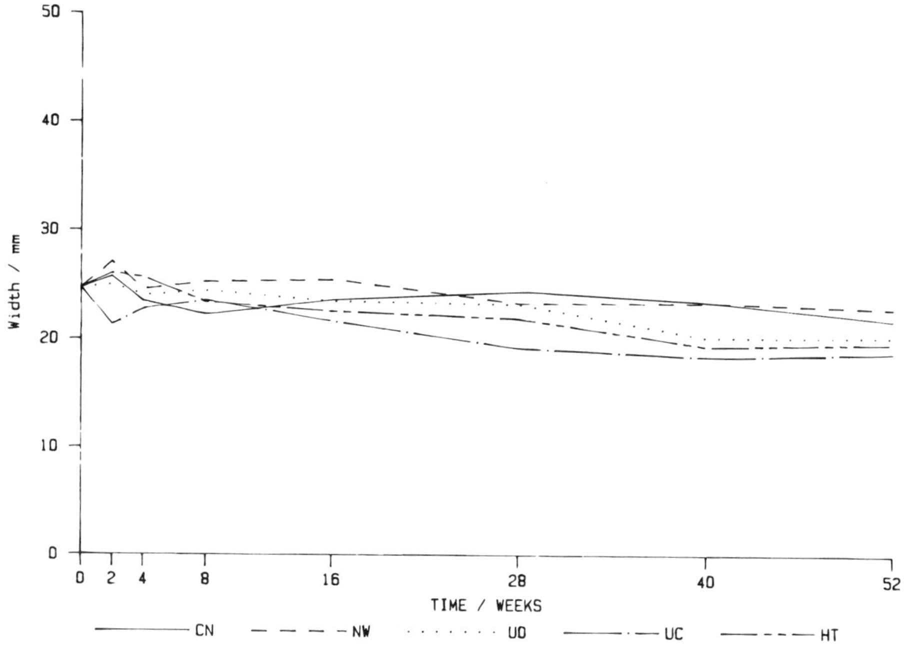

FIG. 9—*S13 Year 2 width at maximum force.*

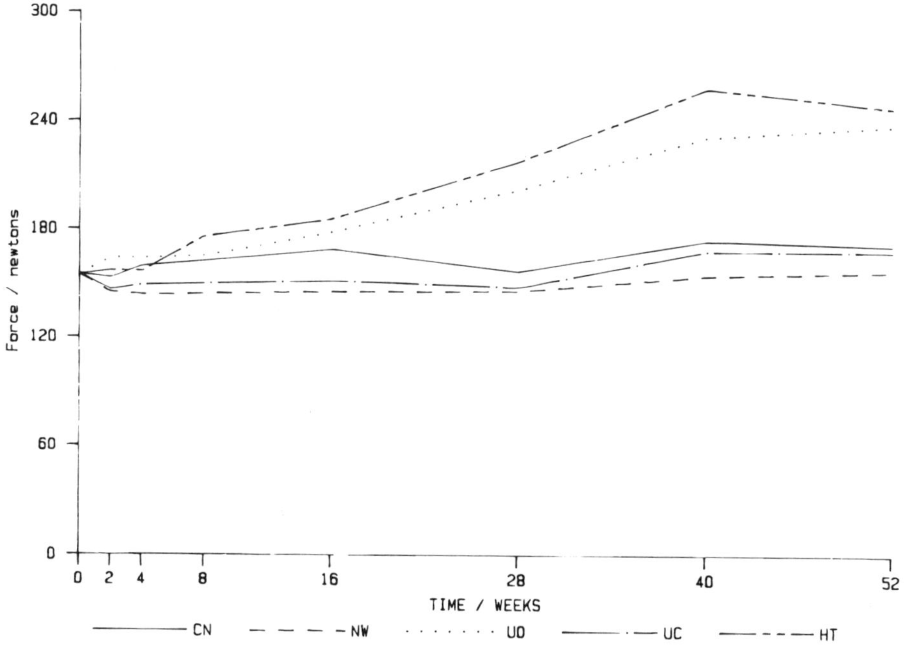

FIG. 10—*S13 Year 2 force at 15 mm width.*

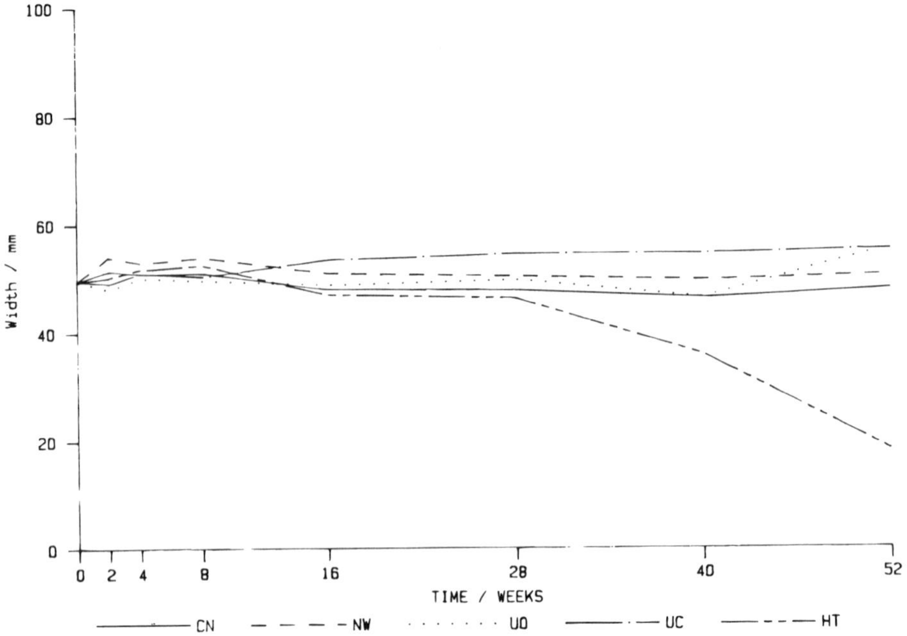

FIG. 11—*U3 Year 2 width at maximum force.*

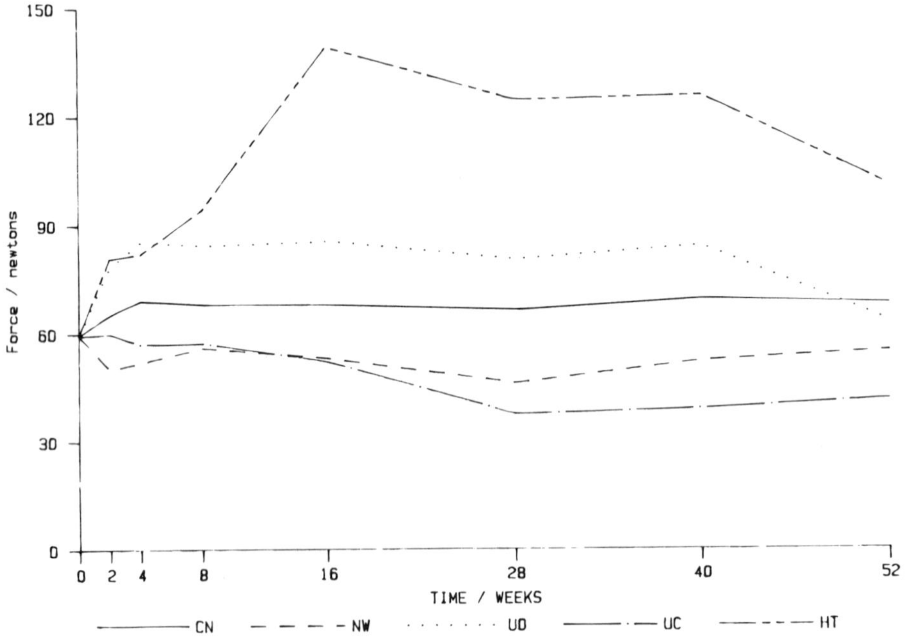

FIG. 12—*U3 Year 2 force at 15 mm width.*

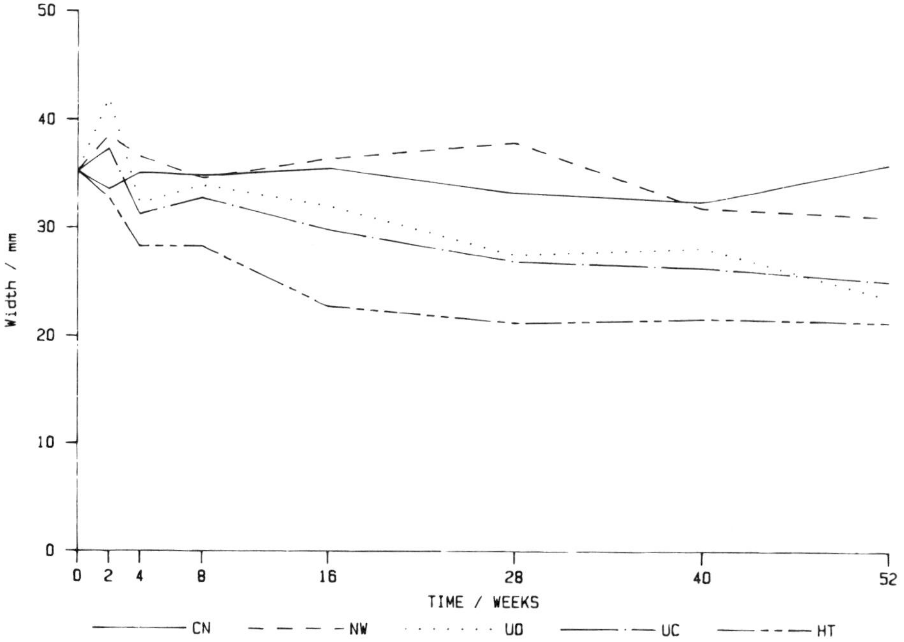

FIG. 13—*U6 Year 2 width at maximum force.*

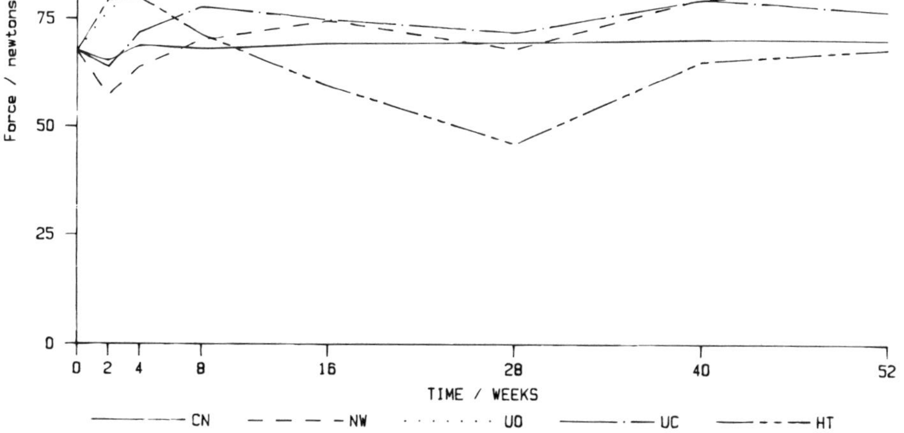

FIG. 14—*U6 Year 2 force at 15 mm width.*

In contrast the laboratory exposures caused more marked changes in modulus and extensibility. Most types of sealants showed the largest increases in modulus over the 52 week period when exposed to heat (70°C). See for example, Fig. 6 for 2P15; Figs. 8 and 10 for two silicone sealants, and Fig. 12 for U3 (a multi component polyurethane). In contrast the one component polysulphide showed modulus values significantly lower than controls for samples exposed to heat and to the other laboratory exposures (Fig. 4).

The polyurethane sealants cannot, however, be treated as a homogeneous group of materials in their response to the degradation factors, owing to the diversity of formulations covered. A comparison of the modulus changes for U3 (Fig. 12) and U6 (Fig. 14) exemplifies this: heat at 70°C (HT) and the UC exposures each produce the opposite effects in these two sealants. As a generalization, the polyurethane materials were all influenced by the laboratory exposures involving exposure to UV(B) radiation: UO and UC. For U3, UC exposure consistently caused reduction in modulus relative to controls. However the maximum reduction in modulus observed after the longer periods of exposure (greater than 28 weeks) in the QUV apparatus was approximately 40 per cent, which is slightly lower than the change observed for the same exposures in the first phase of the study.

It may be recalled that contrasting results from exposing this sealant to six months in the QUV apparatus (using the same test conditions) were reported by Bridgewater and Carbary in a paper [*5*] given at the 1992 ASTM sealants symposium. There is no evidence in this study of the general catastrophic loss of modulus for polyurethane sealants exposed to UV radiation which was found by these authors. In fact the other two polyurethane sealants (U1 and U6) show some tendencies to increased modulus with increasing exposure to UV(B): for example, Fig. 14. The influence of color has been suggested[2] as a possible explanation for the differences in behavior with simulated weathering of this particular polyurethane sealant (U3), found in the two separate experimental studies. This clearly merits further investigation.

Relationship Between Modulus and Extensibility

In general, it may be seen from the graphical data that for those sealants and exposures in which large (and statistically significant) changes take place, there is a consistent near linear relationship between extensibility and modulus. Analysis of the data revealed that for several of the polysulphide and silicone sealants, there was a highly significant negative correlation between changes in extensibility and changes in modulus (the changes in extensibility and modulus from the initial mean values for control samples) during heating at 70°C for up to 52 weeks. This is illustrated in Fig. 15. Each of these correlations is statistically significant at the 0.1 per cent level.

Highly significant correlations were also found between changes in extensibility and changes in modulus in the data (Fig. 16) for exposure of two polyurethane sealants (U1 and U3) to UC and UO cycles. The relationship for U6, not shown in Fig. 16, was significant at the lower level of 5 per cent. In Fig. 16 the previously discussed contrast in behavior between U1 and U3 when exposed to UC cycles is apparent: for U1 both extensibility and modulus decrease with longer durations of exposure; for U3 the expected negative correlation is obtained, with decreasing modulus.

The inverse relationship between extensibility and modulus is what might be expected intuitively though as we have seen it is not invariably found in these experimental data. In the cases of control and naturally weathered sealant samples this may be because the changes in both properties between successive exposure periods are small and not significant. In a

[2] Personal Communication, J. M. Klosowski, 1992.

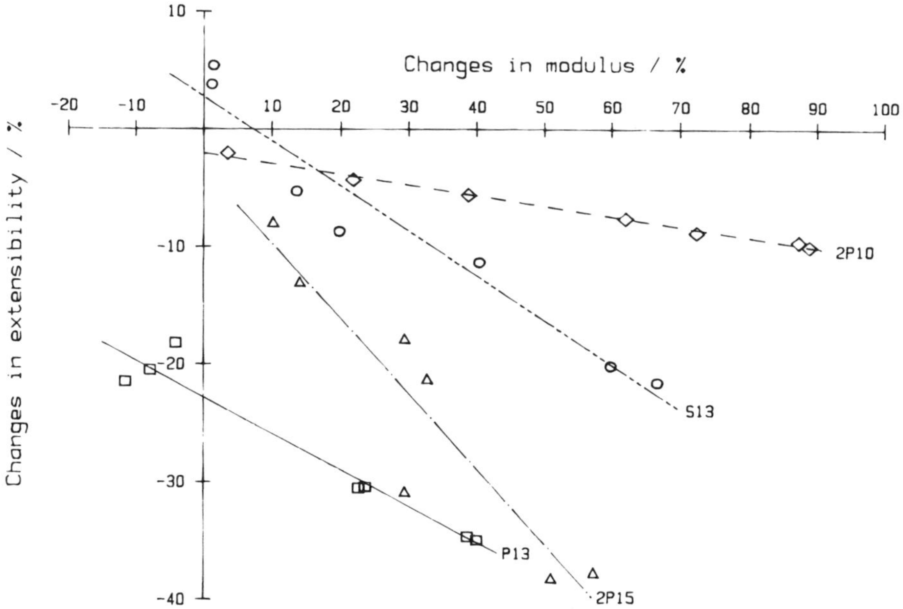

FIG. 15—*Effect of heat (70°C) on extensibility and modulus changes.*

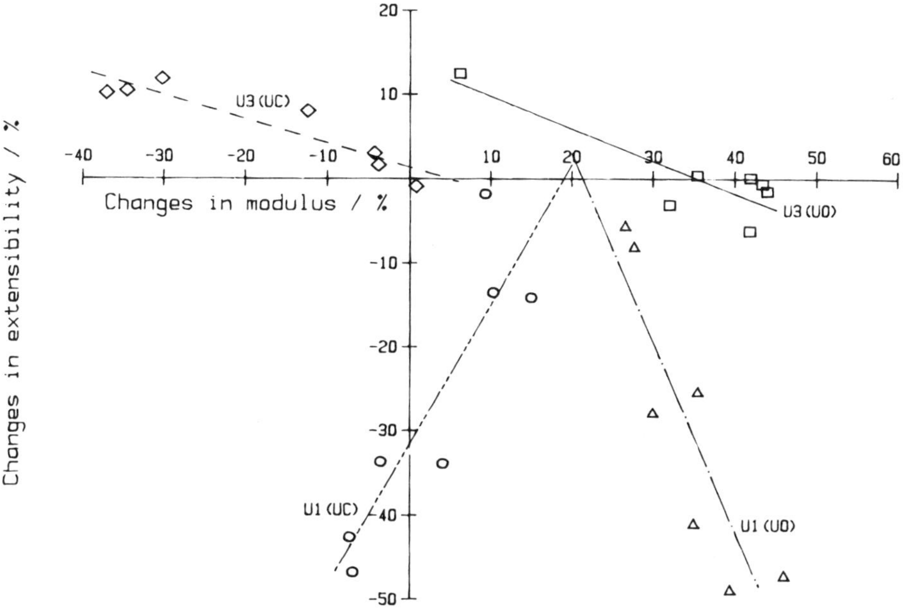

FIG. 16—*Effect of UC and UO on extensibility and modulus changes: U1 & U3.*

number of cases, the sealants exposed to weather show an initial large change in modulus, probably as a result of exposure to condensation. Exposure commenced in spring, when low overnight temperatures often occur with a saturated atmosphere in England. The results of this study suggest that modulus increases may be a sensitive indication of potential loss of durability as a result of exposure to weather or other service factors causing degradation. The consequential reduction in extensibility represents a loss of bond integrity, which is likely to reflect real conditions in a building joint.

It appears possible that the unexpired lifetime of sealant in a building joint might be estimated from a knowledge of the rate of loss of bond integrity obtained from experimental measurements. Such measurements might be complemented by on site measurement of the hardness, which can be related to modulus, of sealants in joints. However hardness measurements such as Shore A are subject to inaccuracy under these conditions.

Further work seems justified to explore relations in changes of extensibility and modulus as a means of durability assessment, in conjunction with various methods of laboratory simulation of weather and service conditions.

Correlations of the Effects of Natural Weather and Artificial Weathering

As has been stated previously, changes in properties during 52 weeks exposure to temperate weather are relatively small for well cured sealants used in this second phase of the study. It is therefore to be expected that only comparatively short durations of exposure to heat, UV radiation, etc., in the laboratory may produce comparable changes in these materials. Following the examination of data for appropriate exposures after the first year [*1*], a comparison for both modulus and extensibility data, was made between the changes after 52 weeks exposure to weather and the effects of the laboratory tests. Generally there were fewer significant correlations than were found in the previous phase of experiments. Two highly significant results were obtained: for extensibility changes, a correlation at 0.1% level, between data for 8 weeks heating at 70°C and for weather (Fig. 17); for modulus changes, the correlation between 8 weeks of UC cycling and weather was significant at the 1% level (Fig. 18). It will be seen [*1*] that this same laboratory exposure yielded a correlation of the modulus values for 52 weeks, at 0.1% level of significance, with the partially cured sealant samples used in the first phase of the study.

Effect of Cure on Response to Natural and Artificial Weather

Much additional analysis of data remains to be completed to compare the results obtained in the first phase of experiments, and those for year II now discussed. For the present, we confine our consideration to two sealants only: the one component polysulphide, only approaching 'plateau' conditions of cure after 2 years storage; and S13, the high modulus silicone, which exhibited the fastest rate of cure in year I and was essentially fully cured through year II. Figures 19 and 20 contrast the changes in these two materials over a two year period, for, respectively, modulus and extensibility. The initial control values at the start of year I are taken as the datum values. The changes resulting from exposure to weather and in the QUV apparatus, for each phase, are superimposed upon the control values. The effects of each of these conditions, relative to the control values, are in general proportionately less in the second year of exposure, for both sealants. The magnitude of the effects is however considerably greater for the polysulphide, suggesting that a relatively uncured sealant is more vulnerable to degradation by weather factors than a well cured sealant. However since S13 and P13 are of entirely different formulations, a firm conclusion on this point is not possible from present data.

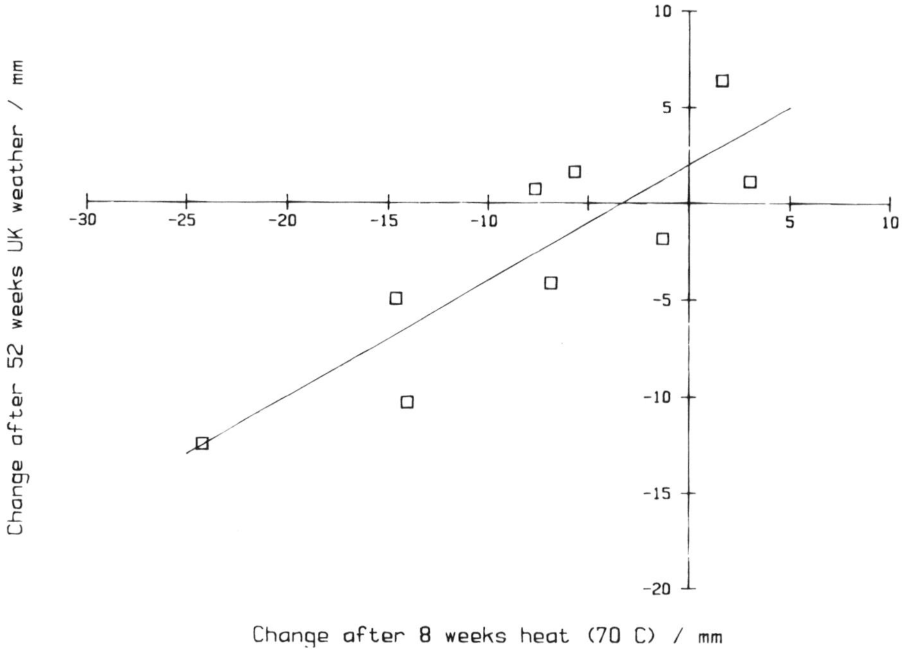

FIG. 17—*Correlation of extensibility change data: heat (8 weeks) with UK weather (52 weeks).*

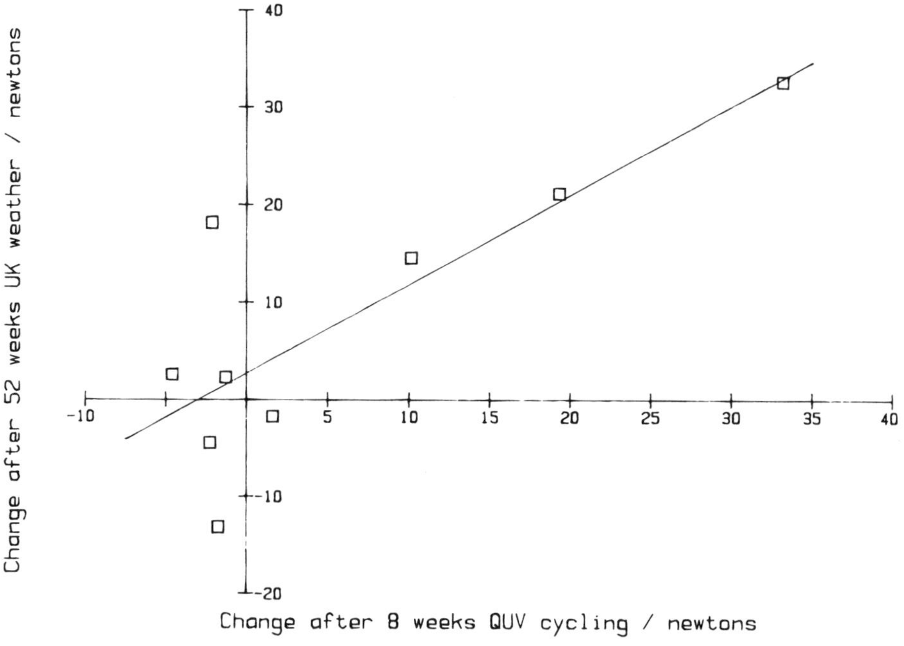

FIG. 18—*Correlation of modulus change data: UC cyles (8 weeks) with UK weather (52 weeks).*

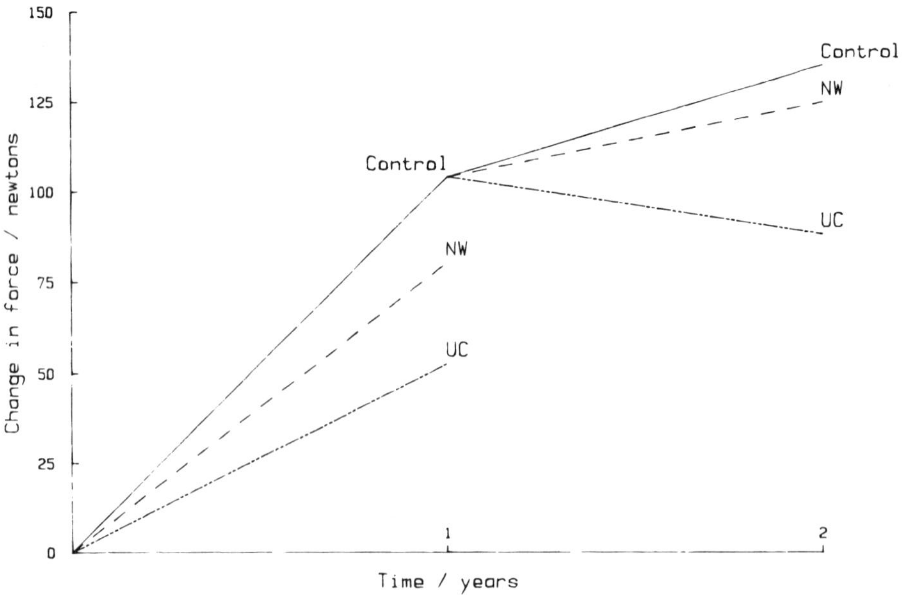

FIG. 19a—*P13: Modulus changes in years 1 & 2.*

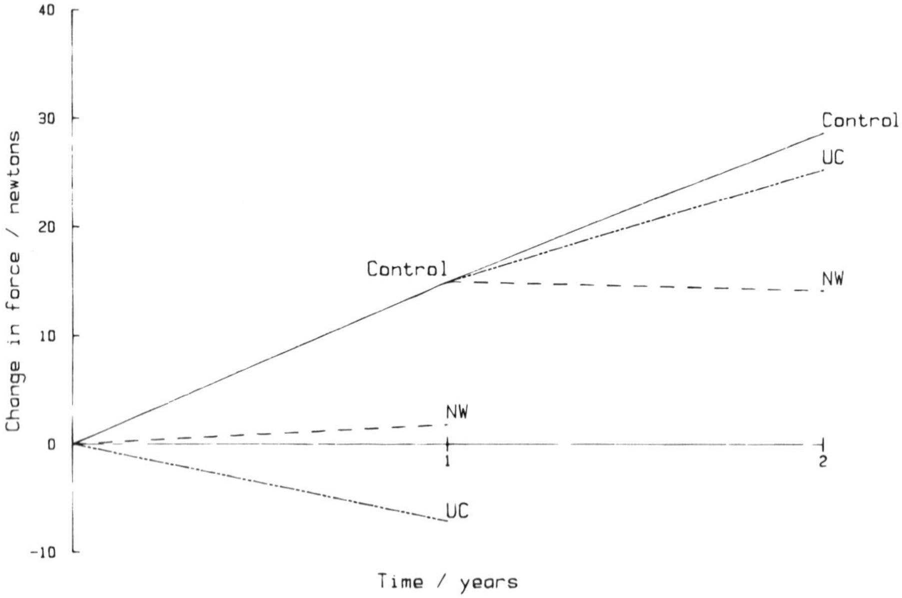

FIG. 19b—*S13: Modulus changes in years 1 & 2.*

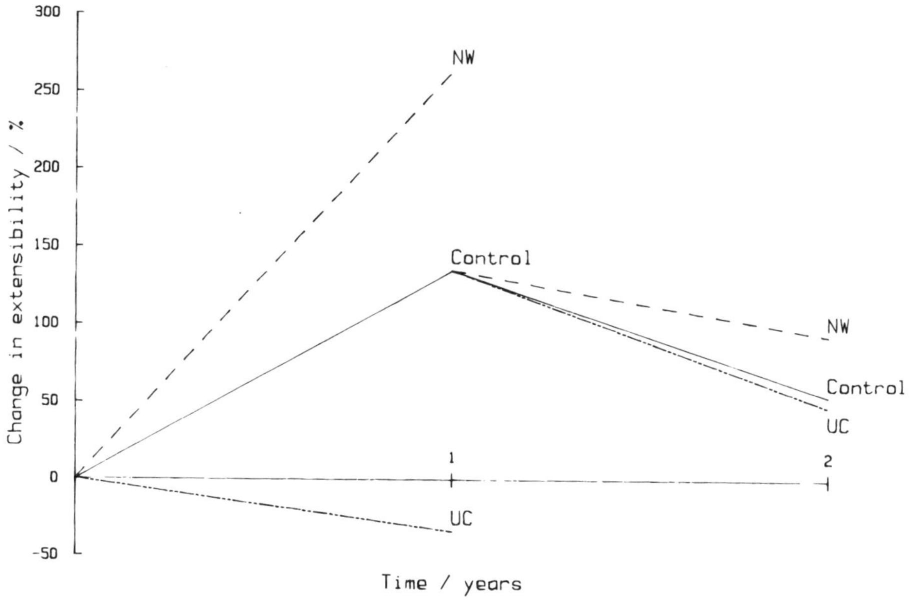

FIG. 20a—*P13: Extensibility changes in years 1 & 2.*

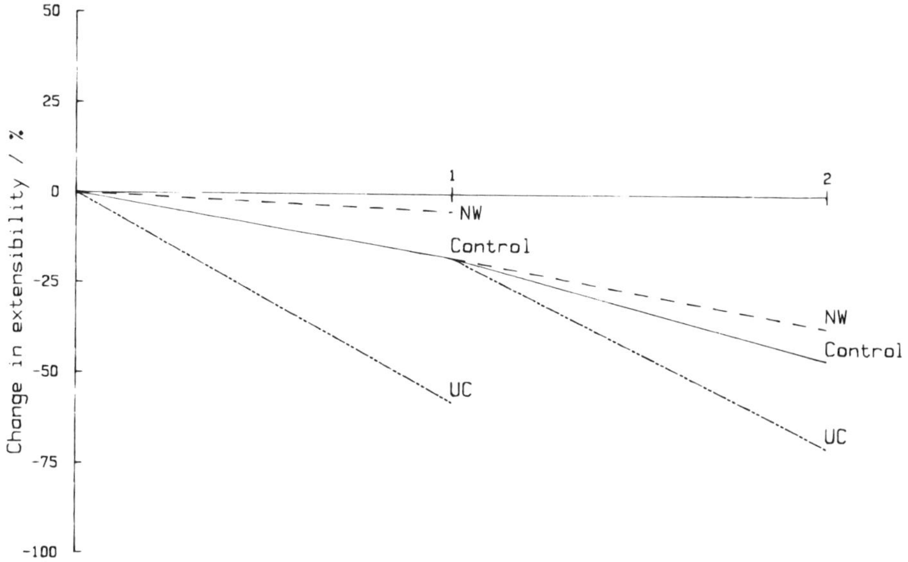

FIG. 20b—*S13: Extensibility changes in years 1 & 2.*

A preliminary analysis of changes in modulus and extensibility from controls indicates some correlation between the data obtained in the two phases of the experiments. Further analysis will be required to elucidate this, and to investigate the influence of cure upon sealant performance.

Conclusions

The majority of the sealants studied had attained stable properties after 12 months storage under control conditions. The changes in properties as a result of exposing one year old sealants samples to natural and artificial weather were therefore more readily discerned than in the first phase of work on freshly cured sealants reported earlier.

Heat aging at 70°C resulted in large and largely consistent changes in modulus and extensibility for polysulphide and silicone sealants. Some highly significant correlations were found between the changes in these two parameters for these materials.

Polyurethane sealants were more affected by laboratory exposures involving UV radiation: UC cycling and UO at high temperature, 70°C. Again changes in modulus correlated significantly with changes in extensibility. These results show some potential for durability assessment based on laboratory tests.

Highly significant correlations were found between the property changes after exposure to weather in England for one year and in certain laboratory tests. Extensibility changes after 8 weeks (1300 hours) heating at 70°C, and modulus changes after 8 weeks of UC cycling thus appear to have some potential for predicting durability.

This requires to be further tested by an examination of data for those sealants products which have been exposed for longer periods, up to 15 years in the cases of both English and Middle Eastern climates.

References

[1] Beech, J. C. and Beasley, J. L., "Evaluation of Cure and Durability Aspects of Sealants," *Science and Technology of Building Seals, Sealants, Glazing and Waterproofing, ASTM STP 1200,* Jerome M. Klowsowski, Ed., American Society for Testing and Materials, Philadelphia, 1992.

[2] British Standard Specification BS5215, 1983, One Part Gun Grade Polysulphide Based Sealants, BSI, London, 1983.

[3] British Standard Specification BS4254, 1986, Two Part Polysulphide Based Sealants, BSI, London, 1986.

[4] British Standard Specification BS5889, 1989, One Part Gun Grade Silicone Based Sealants, BSI, London, 1989.

[5] Bridgewater, T. J. and Carbary, L. D., "Accelerated Weathering and Heat Stability of Various Perimeter Sealants," *Science and Technology of Building Seals, Sealants, Glazing and Waterproofing, ASTM STP 1200,* Jerome M. Klowsowski, Ed., American Society for Testing and Materials, Philadelphia, 1992.

Anthony R. Fiorillo[1]

Case Histories of Successfully Sealed Expansion Joints with Polysulfide Sealants

REFERENCE: Fiorillo, A. R., "**Case Histories of Successfully Sealing Expansion Joints with Polysulfide Sealants,**" *Science and Technology of Building Seals, Sealants, Glazing, and Waterproofing: 3rd Volume, ASTM STP 1254*, James C. Myers, Ed., American Society for Testing and Materials, Philadelphia, 1994, pp. 51–53.

ABSTRACT: Polysulfide polymers have been used as the base polymer in elastomeric sealants for over 40 years. This paper will discuss some of these applications and will show the versatility of these sealants to perform well in various environments.

KEYWORDS: polysulfide, elastomeric, sealant, curtain wall

In the late 1930s, a polysulfide based sealant was used to seal the fuel tanks for Pan American World Airways' "China Clipper." This was the first use of an elastomeric sealant based on a synthetic polymer and is the first in a long list of installations for sealants based on the polysulfide polymer.

With the advent of World War II, the use of polysulfide based sealants expanded to include a number of military applications ranging from sealing fuel tanks in aircraft and tanks to sealing the hulls of ships.

After the war, the suppliers of polysulfide based sealants began looking for other applications for their sealants.

The good resistance of polysulfide based sealants to petroleum products made them a natural to become the standard for virtually all aircraft sealants. Because of their good weathering resistance and elastomeric properties the construction market seemed to offer the next greatest opportunity.

Around this same time, architects began to take a new approach to building construction. Prior to the 1950s, building construction consisted primarily of load bearing, masonry walls where movement was minimal. Oil based caulks were widely used for sealing joints and any other areas in these walls where there was potential for water or air leakage.

In the 1950s curtain wall panels began gaining popularity. The conventional sealants available at that time could not withstand the stresses put on them by the high movement these panels went through with changes in temperature. Elastomeric sealants were needed that would be capable of absorbing these stresses.

Architects began specifying polysulfide based sealants for use in these applications. In many cases these sealants have performed well for periods of 25 years or more.

[1] Group Leader, LP Technical Service and Applications, Morton International, 1275 Lake Ave., Woodstock, IL 60098.

One of the first uses of a polysulfide sealant in a construction application was in 1947 when a polysulfide sealant that had been developed to meet a government specification for sealing the hulls of ships was used to seal the windows in a new California Telephone and Telegraph Company building. That sealant was still performing more than 30 years later.

In 1953 the Lever house in New York City was one of the first high rise buildings to abandon the traditional structural masonry for the attractive curtain wall construction. Thus it was one of the first buildings to experience the problems associated with this new architectural form, significant joint movement.

The building was originally sealed with a typical oil based caulk. Six months after the building was completed, there were serious leakage problems through virtually every joint. The original oil based caulk had to be removed and replaced with an elastomeric sealant that would expand and compress with the movements of the panels. The sealant chosen for the job was a polysulfide based sealant. Much of this original sealant is still performing today, almost 40 years later.

In 1956 the International Teamsters and the Federal Reserve Buildings in Washington, D.C. were sealed with polysulfide sealants. That same year, the Texaco Oil Company built a highrise office building in downtown New Orleans. A polysulfide sealant was used to seal this building, which is now part of the Tulane Medical Center. Again, this original sealant is still performing today, more than 35 years later.

That same year, the Seagram's Building and the United Nations Building in New York City were also sealed with polysulfide based sealants. Much of these original sealants are still in tact.

In 1958 the Pan Am Building in New York, the Federal Reserve Bank in Detroit, the IBM building in Chicago, the Union Oil building in Los Angeles and the Sears Building in Coral Gables, Florida were all sealed with polysulfide sealants.

In 1959 the Air Force Academy in Denver, Colorado and the American Airlines Building at Kennedy Airport in New York were sealed with polysulfide sealants. In 1962 the NASA Complex in Houston, Texas was sealed with a polysulfide sealant.

In 1964 a new Post Office Tower, which extended 580 feet above street level, was built in London. It was designed to contain telephone trunk and television microwave links so it had to be higher than its surroundings in order for the signals to be effective over a maximum distance. This structure dominated the London skyline and so it was completely exposed to wind, rain, sunlight and a general corrosive atmosphere. The Chief Architect's Division of the Ministry of Public Buildings and Works who designed the tower specified polysulfide based sealants for the areas which would be subjected to the most severe environmental conditions, the external joints and bedding of pressed curtain wall units and windows on the top six floors of the tower. This original sealant is still performing today, almost 30 years later.

In 1967, the Munson-Williams-Proctor Institute in Utica, New York was resealed with a polysulfide sealant. These joints are still water tight and performing well. Around this same time, Saint Joseph's Cathedral in Hartford, Connecticut was built and sealed with a polysulfide based sealant.

A number of the buildings at West Point Military Academy on the Hudson River in New York have been resealed with polysulfide based sealants. The highly impressive gothic chapel at the Point which was constructed of granite in 1904 was resealed with a polysulfide sealant about this same time.

In 1974 the Sears Tower, the world's largest private office complex and the world's tallest building, containing 110 stories was built. The frame of the building is clad with black aluminum curtain walls consisting of aluminum covers, spandrel panels, extrusions and over 16 000 bronze-tinted, glare-reducing windows. To provide a seal against the adverse conditions of the Chicago weather, a polysulfide sealant was used to seal the joints between the window components. Unfortunately, there have been a number of non-sealant related problems with this building and so most of the original sealant has been replaced.

In the late 1960s and early 1970s a number of buildings and complexes were built in South Africa and sealed with polysulfide sealants. These included the Hillstar Flats and Parkade in Durban, which contained a parking garage, department store and an apartment house (1967). The Elizabeth Hotel (1969), The Carlton Center (1970) and The Sanlan Center (1971), the main building at the University of Port Elizabeth (1972), Natal Provincial Administration Headquarters, Medical Chambers of Natal (1972), many of the buildings at The University of South Africa, also known as UNISA (1972), and the Civil Engineering Building, to name a few. In many cases the sealants used in these installations were not only subjected to large changes in temperature but they were also subjected to high humidity and the corrosive atmosphere found near the ocean.

All of these buildings still have the original sealant in them and although there is some discoloration and surface cracking evident, in every case the sealant is functioning well. Several of these buildings have had additions put onto them and in every case the addition was sealed with a polysulfide sealant.

In 1968, Tokyo's first high rise, the Kasumigaseki building, was built. Because of the frequency of earthquakes in Japan, this building was designed to withstand earthquakes with a magnitude of 7.1 on the Richter scale. After numerous earthquakes the polysulfide sealant used to seal this building was still performing well more than 20 years later. Ten years later, in 1978, Japan's tallest high-rise, the Sunshine City building was built and also sealed with a polysulfide sealant.

Based on the success of polysulfide based sealants in these applications, architects began specifying them for many construction applications where there was the potential for significant joint movement.

In 1964 the State of California, Department of Water Resources, issued specification CL-2, which gives test methods for "Cold Applied Two Component Modified Polysulfide Polymer Type Joint Sealing Compounds" for sealants used in expansion joints in their aqueduct system. In 1966 the Federal Government approved a polysulfide based sealant for sealing the Federal Government's portion of the California Aqueduct System.

In 1983 a polysulfide based sealant was approved by the state of Arizona for use in their Central Canal project.

Today polysulfide sealants are still being supplied for both, the California Aqueduct and the Central Arizona Canal Project.

Polysulfide sealants were the first high performance, elastomeric sealants to be used in building construction. Because they have been around for over 40 years they are often referred to as "old technology." Applicators sometimes feel that the "new generation" of high performance sealants, the silicones and urethanes, being newer materials, must be improvements over the polysulfides and so today polysulfide sealants are often being replaced with silicones and urethanes.

Because a material has been around for a long time does not mean that it has outgrown its usefulness. These few success stories show that polysulfide based sealants, properly formulated and applied, will give long term protection under various environmental conditions.

Acknowledgment

I would like to express my gratitude to Carl Harvath, Technical Support Manager of Koch Materials; and Mike Finn, Technical Service Manager of Morton International for their assistance in gathering the data presented in this paper.

Reference

Materially Speaking - Morton International.

Design

Dean A. Rutila[1] *and Stephen S. Ruggiero*[1]

Expansion Joints in Deck Waterproofing Systems

REFERENCE: Rutila, D. A. and Ruggiero, S. S., "**Expansion Joints in Deck Waterproofing Systems,**" *Science and Technology of Building Seals, Sealants, Glazing, and Waterproofing: 3rd Volume, ASTM STP 1254*, James C. Myers, Ed., American Society for Testing and Materials, Philadelphia, 1994, pp. 57–70.

ABSTRACT: Structural expansion joints that interrupt plaza and traffic bearing waterproofing membranes are often a source of leakage. Most joint seals in plaza type waterproofing systems are "buried," making them difficult and expensive to maintain. Exposed joint seals in traffic bearing waterproofing systems are subject to severe loading conditions, such as vehicular traffic, but are generally accessible.

This paper presents design concepts for waterproofing expansion joints in building decks. We review assemblies commonly used to waterproof expansion joints and we examine several design concepts. Membrane/waterproofing continuity across joints and joint maintenance are of particular concern. An estimate of the joint movement is necessary to select a joint seal. Adequate drainage away from the joints remains the most important design consideration

KEYWORDS: expansion joints, joint seals, waterproofing, plaza decks, parking garage, building decks, drainage

Expansion joints through plaza building decks must have seals that are compatible with the waterproofing system, overburden and wearing surface; the seals must also accommodate structural movement and pedestrian or vehicle traffic. Most expansion joint seals are buried and protected by the many layers in a plaza system consisting of waterproofing, drainage, insulation and wearing surface components. Some expansion joint seals are exposed and subjected to traffic, such as in parking garages or similar plazas that have only a waterproofing membrane and wearing surface, or an integral waterproofing/wearing surface.

Based on our experience reviewing non-performing installations, seals for plaza expansion joints are plagued by a lack of drainage away from the joints. Drainage problems exist at joint seals, even given long published recommendations to provide drainage, for example, ASTM C 898 Guide for Use of High Solids Content, Cold Liquid-Applied Elastomeric Waterproofing Membrane with Separate Wearing Course. Most joints have at least one connection between a waterproofing component and the joint seal, which is readily attacked by ponding water.

Problems with leakage through plaza expansion joints have been so common and difficult to repair that gutters have been installed below expansion joints in new waterproofing installations. Manufacturers continue to develop new seal technology/components as some earlier products prove less than perfect. Twenty-five years ago, a few manufacturers produced waterproofing membranes and a few manufacturers produced expansion joint seals for roofing systems; seals for waterproofing plaza decks were not marketed until about 1970 when gaskets were manufactured for the purpose of supporting sheet membranes over expansion joints [*1*]. Ten

[1]Professional Engineers, Simpson Gumpertz & Heger Inc., Consulting Engineers, Arlington, MA.

years ago, the joint seals now available were marketed in one form or another, but most of these have continued to "evolve." Recent marketing appears to concentrate on a few "hot" product types with more producers each year. At the same time, the roofing expansion joint seal from 25 years ago is little changed and is still an option for plaza joints.

Types of Expansion Joint Seals

Figures 1 and 2 show the various expansion joint assemblies that are commonly used in deck waterproofing applications. Seals for building deck expansion joints are generally achieved either by maintaining continuity of the waterproofing membrane across the joint (Detail A), or by connecting the waterproofing membrane, in some fashion, to a discrete joint assembly that accommodates movement while providing a waterproofing seal (Details B through E). The following is a review of the salient features of the various joint assemblies shown in Figs. 1 and 2.

Continuous Waterproofing Membrane

Loose-laid, single-ply waterproofing membranes can be installed continuously over the expansion joint, if given adequate support and geometric flexibility to accommodate movement.

1. Certain thermoplastic (PVC) and elastomeric (butyl) membranes perform well in this type of assembly when installed over a preformed bellows expansion joint cover. The bellows provides support for the membrane and assures proper membrane geometry to accommodate movement.
2. The most critical factor in the joint construction is achieving watertight transverse seams between membrane sheets at locations along the length of the joint. Care must be taken with the adhesive seams in elastomeric sheet membranes, to ensure adequate pressure is applied during the seaming process without distorting the geometry of the bellows.

Prefabricated Bellows

The typical prefabricated bellows used in roofing expansion joint assemblies, for many years, may also be used to waterproof building deck expansion joints (Fig. 1B). This type of joint cover is typically fabricated with a neoprene or EPDM bellows bonded in the factory to metal (stainless steel or copper) flanges.

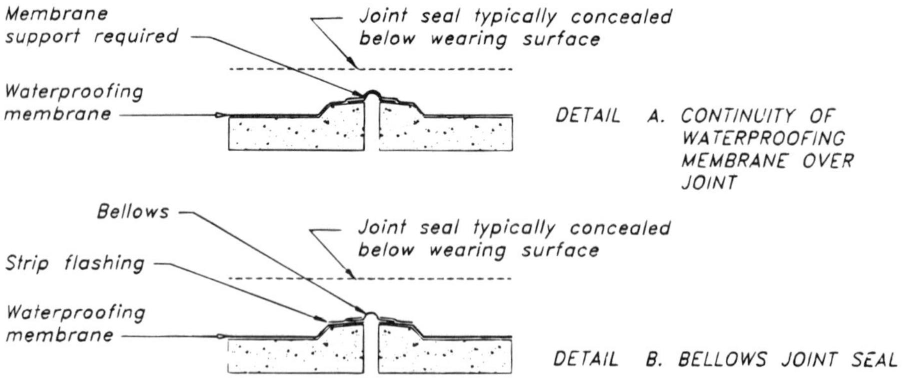

FIG. 1—*Types of seals for concealed expansion joints in deck waterproofing systems.*

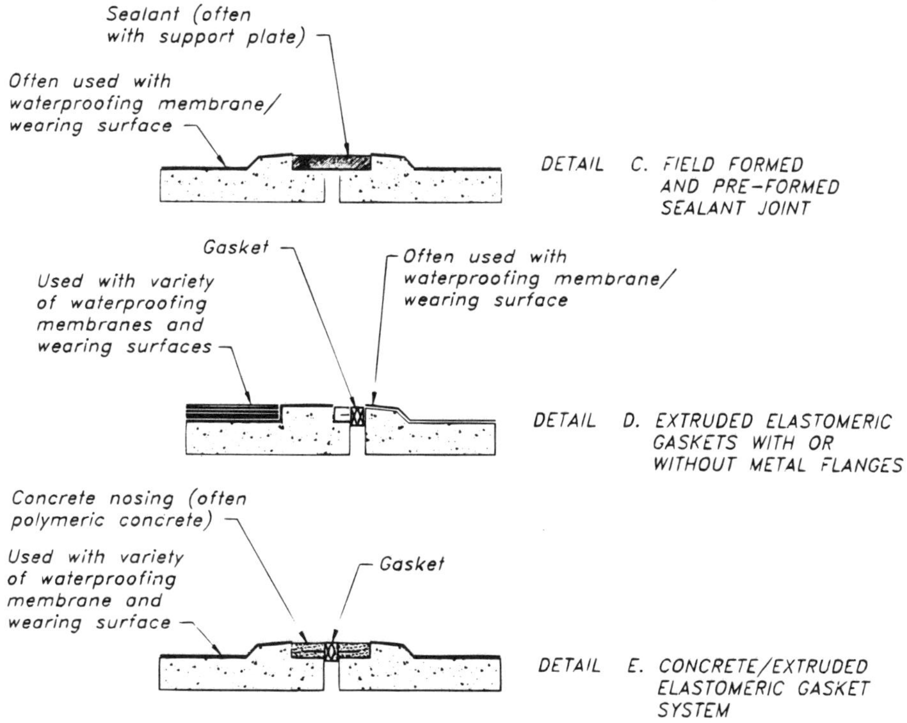

FIG. 2—*Types of seals for exposed expansion joints in deck waterproofing systems.*

1. Factory fabricated corners and transitions are readily available accessories that are essential to achieving a waterproof joint.
2. The authors have found that many of the membrane manufacturers have omitted this type of joint seal from their system details. With proper detailing of the transition between the waterproofing and the metal flanges of the joint cover, we find that this type of joint assembly can provide a good expansion joint seal for any of the various waterproofing systems (loose-laid, fully adhered or liquid-applied).

Field Formed or Pre-formed Sealants

Fluid-applied sealants and pre-formed sealant gaskets are often used to seal the surface level of expansion joints (Fig. 1C). Their relatively low-cost makes them a common choice for exposed applications such as parking garages.

1. Sealant applied between concrete slab edges, metal shoulders in slab edges, or similar vertical surfaces are inexpensive and easily maintained, but not sufficiently reliable for most plaza decks located over inhabited space. The suitability of these joints for pedestrian or vehicle traffic should be considered during design.
2. Sealant applied over a recessed support plate better accommodates vehicular traffic than a butt joint sealant but is only marginally more reliable. The deck recess is generally wider than the sealant, and is filled with a nosing concrete.
3. Continuity of the waterproofing relies on the bondline at either edge of the sealant.

Extruded Elastomeric Gaskets

Vulcanized and non-vulcanized elastomers are extruded into matrix shapes with large closed cells, solid shapes, or shapes with cellular and solid components. Extruded shapes are used with and without metal support components to provide surface seals for expansion joints.

1. Cellular compression gaskets are installed partially compressed between slab edges, metal shoulders in slab edges, or similar vertical surfaces. Continuity with deck waterproofing is generally impractical. These gaskets are not sufficiently reliable for most plaza deck applications, but are often used in parking garages.
2. Gasket sections are jointed by adhesives or heat welding; heat welded intersections are generally more reliable.
3. Metal shoulders and receivers with butt joints between sections are prone to leakage at the butt joint.

Concrete/Extruded Elastomeric Gasket Systems

Concrete produced with elastomeric or polymeric binders in place of portland cement are combined with extruded vulcanized or non-vulcanized gaskets to provide composite expansion joint seals. Metal support components are a part of some systems.

1. Solid gaskets or cellular gaskets with solid flanges are installed in a recess along the joint using a nosing concrete as adhesive. The elastomeric or polymeric concrete provides waterproofing continuity at the gasket joint seal and must be joined with the deck waterproofing.
2. Heat welded gasket intersections are generally more reliable than sections jointed by adhesives.
3. Certain concrete/gasket systems provide considerable reliability for parking garage or similar plaza decks, but are not sufficiently reliable for most occupied building spaces, because of the difficulty in providing continuity with the waterproofing membrane.

Design Parameters

The following parameters should be considered in the design/selection of waterproof expansion joint seals:

Continuity—between the joint seal and waterproofing components including waterproof seams or connections between components.

Compatibility—of the joint seal with the waterproofing membrane and plaza system. The design of the waterproofing system necessarily impacts the design of the joint.

Movement—of the joint due to thermal cycles of the structure and overburden, shrinkage and creep of concrete decks, and live loads such as vehicular traffic.

Drainage—away from the joint seal, particularly seams or connections between components and the waterproofing membrane.

Protection—for the joint seal, either by a wearing surface or by the intrinsic durability of the joint seal.

Provision for maintenance—to accommodate the greater need for repairs at expansion joints compared to the plaza in general, particularly where surface seals are part of the joint construction.

Continuity of Waterproofing

The waterproofing must be continuous over the expansion joint, otherwise a hole is literally built into the system. Providing continuity for the waterproofing is relatively simple with certain sheet waterproofing membranes that can be carried over a supporting bellows (Figs. 3 and 4) or gasket ASTM C 898. In addition to providing a good "seal" for the expansion joint, these two details provide for prompt drainage away from the expansion joint; Fig. 4 admits to movement at the wearing surface level as well as at the building deck level.

Most waterproofing systems require several components, often by different manufacturers, to properly seal an expansion joint. Figure 5 provides reasonably good waterproofing continuity

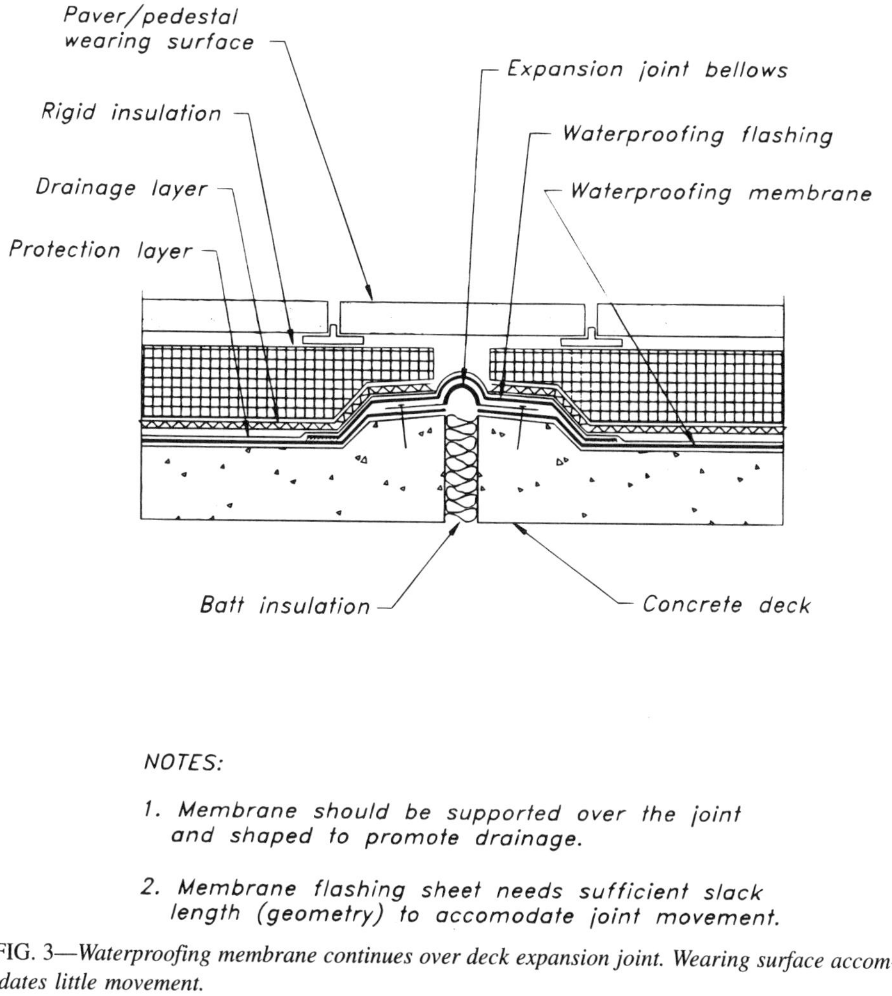

FIG. 3—*Waterproofing membrane continues over deck expansion joint. Wearing surface accommodates little movement.*

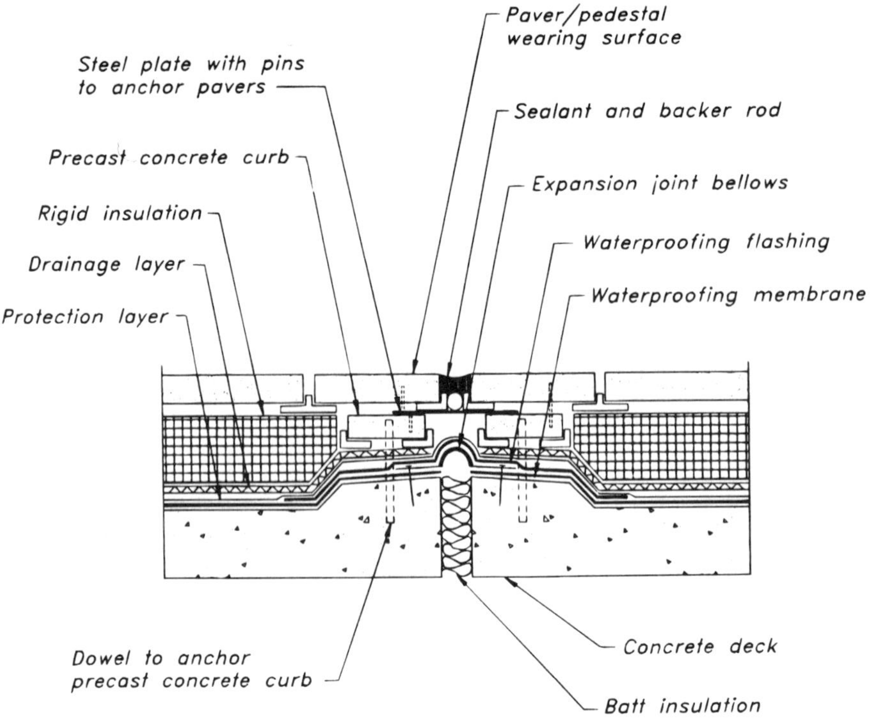

FIG. 4—*Waterproofing membrane continues over deck expansion joint. Wearing surface accommodates movement.*

for adhered waterproofing membranes, with emphasis again on prompt drainage away from the joint. We have used this with modified bituminous/polyethylene laminate sheet systems, but it can also be used with most liquid-applied membranes. Proper control of the concrete finishing is necessary to provide a suitably even surface for fastening the bellows flange and fasteners must be closely spaced, otherwise the flange may warp or "oil can."

The detail in Fig. 5 uses four waterproofing components: sheet waterproofing and two-part liquid-applied waterproofing by one manufacturer, and a metal flanged rubber bellows with sheet rubber seals (strip flashing) for transverse joints by another manufacturer. Particular care is needed at transverse joints to deal with the possible incompatibility of the sheet rubber and modified bitumen of the waterproofing membrane, which are made by separate manufacturers. In this example, the two-part liquid-applied waterproofing separates the two materials.

Figure 6 shows an unsuccessful detail with lack of continuity of the waterproofing as the major flaw. The waterproofing stops at the top edge of the expansion joint curb, where it is particularly vulnerable to leakage. The horizontal surface of the curb is unprotected by waterproofing, so water leaks through cracks in the curb. Also, water leaks at the shoulder angles of the joint because the angles are installed in 10 ft lengths with a small gap (butt joint) between angles.

Figure 7 addresses the major continuity flaws in Fig. 6 by providing a more reliable membrane termination, waterproofing over the curb and a joint seal that is made continuous with the curb waterproofing. Even as modified from Fig. 6, the detail in Fig. 7 is not suited for use over occupied space; however, in parking garage applications it provides waterproofing

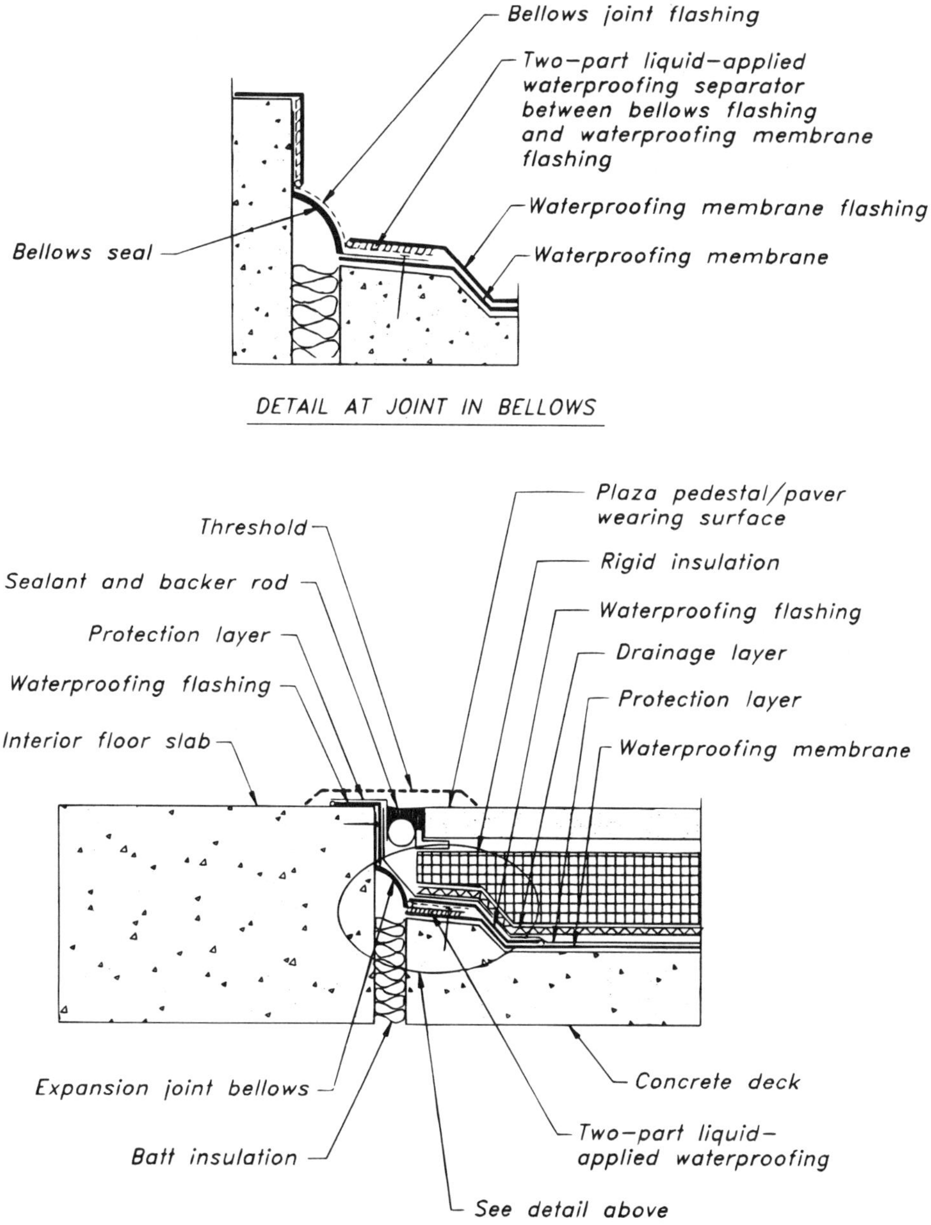

FIG. 5—*Bellows seal over deck expansion joint.*

satisfactory to many owners. Problems include: The lack of continuity between the sheet waterproofing and the curb waterproofing, the lack of drainage inherent in the bituminous paving system, and the difficulty in maintaining (repairing) the sheet waterproofing termination. Also, occupied spaces generally require insulation and Fig. 7 does not accommodate insulation

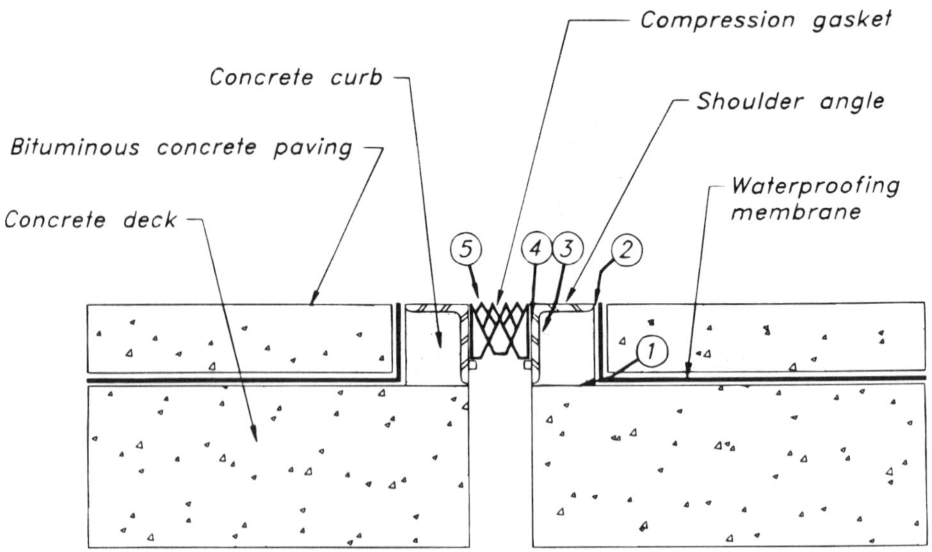

KEY TO LEAKAGE NOTES:

1. Leakage through seams in waterproofing membrane at base of curb, then through curb construction joint.
2. Leakage between curb and waterproofing, then through curb construction joint.
3. Leakage between lengths of shoulder angles.
4. Leakage between gasket and shoulder angle.
5. Leakage through transverse joints in gasket.

FIG. 6—*Leakage paths through expansion joints that lack membrane continuity.*

placement over the membrane. The inset details in Fig. 7 attempt to provide continuity of the waterproofing through the use of liquid-applied sealants.

Estimating Joint Movement

The width of the expansion joint through a new structure is generally established by the architect with advice from the structural engineer. Thermal cycles and inelastic shortening of the concrete deck due to shrinkage and, possibly, creep account for most horizontal joint movement in a typical at-grade plaza structure. The width required by the waterproofing seal, however, may be greater than that determined by purely structural considerations.

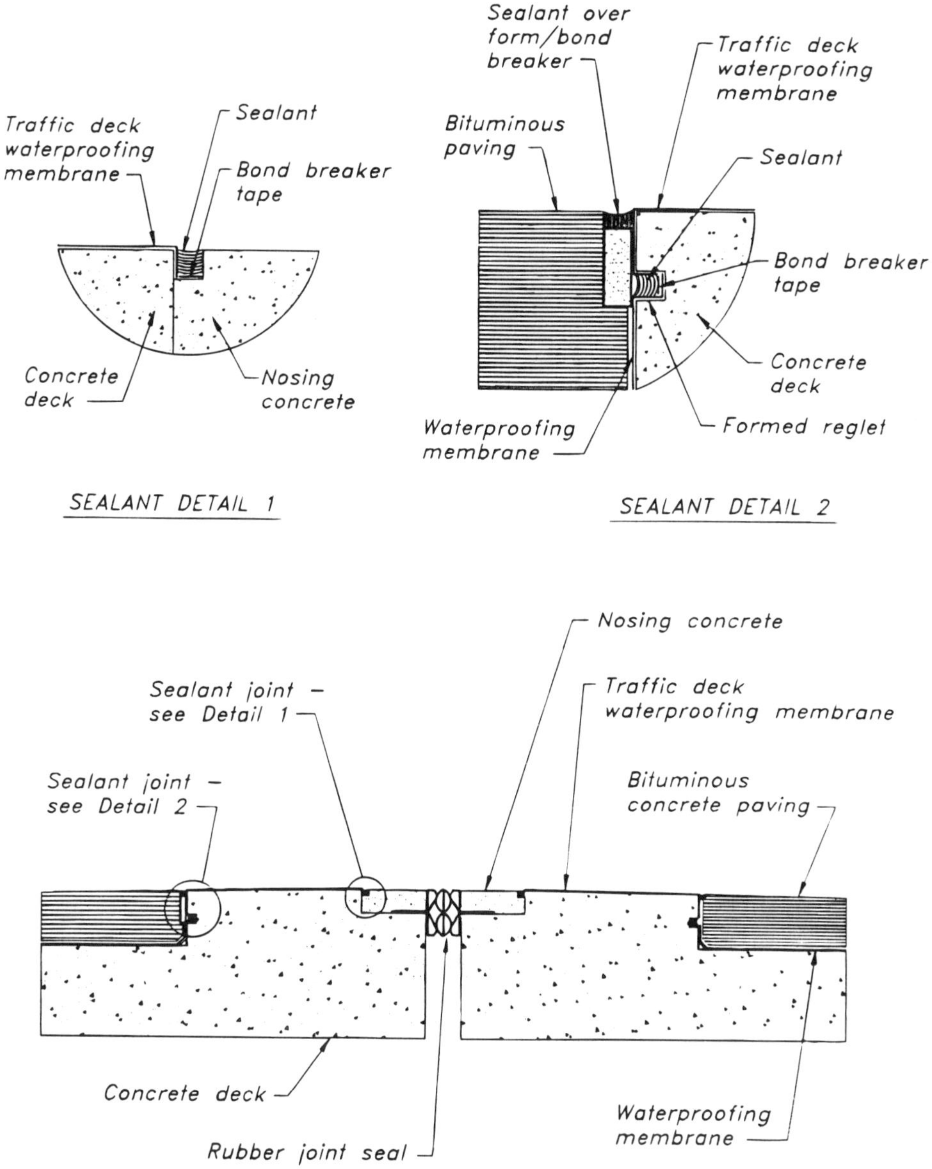

FIG. 7—*Surface seal design for deck expansion joint that accommodates concealed waterproofing. Maintenance of deck waterproofing is a problem.*

A quick estimate of the horizontal movement of a joint is useful when developing a seal design. As shown in the following example, some basic knowledge of the structure is required. A relatively simple frame analysis by a structural engineer will refine an estimate, probably reducing the thermal and shrinkage estimate because of the restraint of the frame. However, in most designs, further refinement is not useful. Joint movement due to deflection of the deck as vehicular traffic crosses the joint, or horizontal movement due to wind loads (or other live loads) should also be estimated by a structural engineer.

Horizontal Movement Perpendicular to the Joint

Figure 8 shows an expansion joint between two plaza areas, each 45m × 45m. We will estimate the joint movement at Location A due to shrinkage and thermal cycles of the concrete deck [2], as follows:

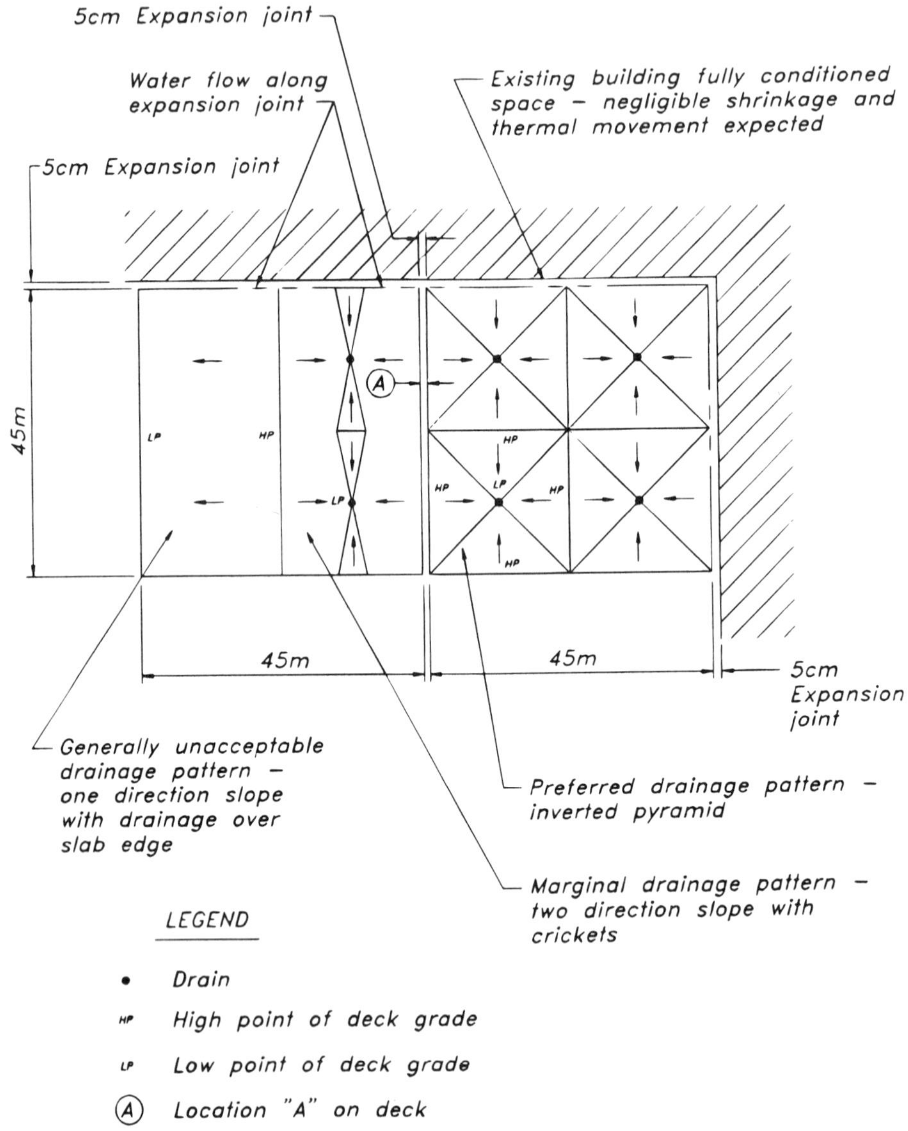

FIG. 8—*Plan layout of deck expansion joints and drainage grades.*

Shrinkage

Assume the waterproofing is installed 60 days after a 7 day moist cure. Estimate shrinkage strain ($\in$) using $\in sh = \frac{t}{(35 + t)} (\in sh)u$ with t = days and $(\in sh)u = 780 \times 10^{-6}$ m/m (in./in.). Assume the shrinkage is unrestrained by the beams and columns. Assume the geometric centers of the two areas remain stationary. The joint movement after the waterproofing is installed is estimated as the difference between the total shrinkage and the shrinkage after 60 days. Estimate the shrinkage movement of a slab as $\Delta = L \in sh$, with L = the perpendicular distance from the joint to the geometric center.

$\Delta sh60$ = deck shrinkage 60 days after curing = $L \in sh$

$$= 22.5 \text{ m} \left(\frac{60}{35 + 60}\right) 780 \times 10^{-6} \text{ m/m} = 0.011 \text{ m} = 1.1 \text{ cm (.44 in)}$$

Δsh total = deck shrinkage total = $L (\in sh)u = 22.5 \text{ m} \times 780 \times 10^{-6}$ m/m
$= 0.018$ m = 1.8 cm (.71 in)

Δsh = joint movement due to shrinkage = Δsh total − $\Delta sh60$
= 0.7 cm (.27 in) after waterproofing installation.

Thermal

Assume that the daily average temperature variation expected from the 10°C (50°F) temperature at the time of installation is a 35°C (63°F) increase and 25°C (45°F) decrease, that is, a temperature range of 45°C to −15°C (113°F to 5°F). Assume the thermal movement is unrestrained by the beams and columns. Assume the geometric center remains stationary. Estimate the movement $\Delta t = Lt\in$, with L as defined above, t = temperature change, and $\in = 10.0 \times 10^{-6}$/°C ($5.5 \times 10^{-6}$/°F).

$\Delta t + 35$ = deck expansion for 35°C increase = $Lt\in = 22.5 \text{ m} \times 35°\text{C} \times 10.0 \times 10^{-6}$/°C
= .0079 m = 0.79 cm (.31 in)

$\Delta t - 25$ = deck construction for 25°C decrease = $Lt\in = 22.5 \text{ m} \times 25°\text{C} \times 10.0 \times 10^{-6}$/°C
= 0.0056 m = 0.56 cm (.21 in)

Total Joint Movement

The deck expansion or contraction computed above is for one slab at the Location A, the movement is doubled because both slabs move. After about 1 year, with most shrinkage complete, the joint width is: $W = 5 \text{ cm} + 2\Delta sh + 2\Delta t25 = 5 \text{ cm} + 2 \times 0.9 \text{ cm} + 2 \times 0.6 \text{ cm} = 7.6$ cm (3 in).

Shortly after construction, before drying shrinkage, at a 45°C (113°F), the joint width is: $W = 5 \text{ cm} - 2\Delta + 35 = 5 \text{ cm} - 2 \times 0.8 \text{ cm} = 3.4$ cm

For this joint, an estimate of the movement that the 5 cm joint seal must accommodate is + 2.6 cm and as much as − 1.6 cm (+52%, −32%).

Design to Accommodate Joint Movement

The design of the joint seal should consider the total range of structural movement, which inevitably depends on the temperature when the joint seal is installed, and the cumulative effect of drying shrinkage. As the example demonstrates, expansion joints in new buildings may experience more movement opening the joint than closing the joint. A seal installed at peak temperatures in the summer may never be narrower and the joint will only get larger. This can be a particular problem for seals that rely on compression gaskets.

The 5 cm expansion joint width in the example may satisfy the architectural and structural considerations, but the joint is too narrow for many seals to accommodate +2.6 cm and −1.6 cm movement. Either the joint width should be increased, or the available options for the seal are reduced considerably.

Drainage Away from the Joint Seal

The building deck must be graded to provide positive slope to drains and to place the expansion joints at the high points. In addition, the joint seal should be raised further above the deck on a curb to promote drainage away from the connections between system components.

The inverted pyramid drainage grid shown in Fig. 8 provides proper slope of the structural deck away from expansion joints to "single point" internal drains. Slope parallel to the joint is less reliable even with a raised curb because flashing connections remain in the drainage flow.

Wearing surfaces with closed joints or no joints must slope away from the expansion joint; otherwise, the joint seal may be exposed to substantial water flows. Closed joint systems include brick in sand setting beds. Jointless wearing surfaces include bituminous paving, concrete wearing slabs, or waterproofing membranes that also act as the wearing surface (deck coatings).

Prompt drainage at the membrane level is one of the most important characteristics of a successful waterproofing installation; it is equally important to the durability of expansion joint seals. Without membrane level drainage in a typical buried waterproofing system, the expansion joint seal and its connections to the waterproofing system remain exposed to water, which may lead to premature failure of the joint materials or their connection to other weatherproofing components. All wearing surfaces should be designed so as to accommodate a true drainage layer at the membrane level consisting of either a prefabricated drainage panel or smooth aggregate.

Examples of drainage at expansion joints.

Figures 3 and 5 show expansion joint designs in a typical waterproofing system with an open joint wearing surface. The concrete deck is constructed to provide a curb monolithic with the deck, avoiding a construction joint at the curb. Milled lumber, preservative treated, is used as an alternate to the concrete curb when loose-laid sheet waterproofing is used. Lumber is less durable because it may decay, but it simplifies construction by permitting the easy nailing of the bellows flange.

Figure 6 shows an unsuccessful expansion joint design, in part due to the lack of membrane level drainage bituminous wearing surface. The system lacks free drainage at the membrane, resulting in a reservoir of water that leaks through the joint at various defects cited previously.

Design for Maintenance

Joint movement, the number of connections between waterproofing components, and traffic loads on surface seals make expansion joints more likely to require maintenance compared to other plaza areas. Where joints are available for inspection and repair, leakage becomes a relatively minor problem should it occur. Joints that are difficult or impractical to repair tend to remain untreated until major structural damage occurs or general plaza waterproofing replacement is required; gutters are often installed below these joints to provide a holding action, but allow structural damage to continue.

Figure 2 shows an expansion joint where the open-jointed wearing surface and waterproofing system overburden can be disassembled for maintenance. Likewise, Fig. 9 shows a design that provides for joint maintenance in a "monolithic" wearing surface consisting of bituminous paving. The concrete apron in Fig. 9 can be removed, but dowels through the membrane are required to anchor the apron; we accept this tradeoff because of the free drainage at the membrane.

Conclusions

The required level of protection against leakage should be established with the building owner when selecting and detailing an expansion joint seal. When a high level of protection is required for an expansion joint seal, the method of sealing should be a major consideration during the selection of the plaza waterproofing and overburden systems. When less protection is required, based on the building use and the occupants tolerance for joint leakage, the expansion joint seal is not as critical to the selection of the waterproofing and overburden.

Sheet waterproofing systems installed over a supporting bellows generally provide the greatest protection against expansion joint leakage (Figs. 1 and 2). The system requires open

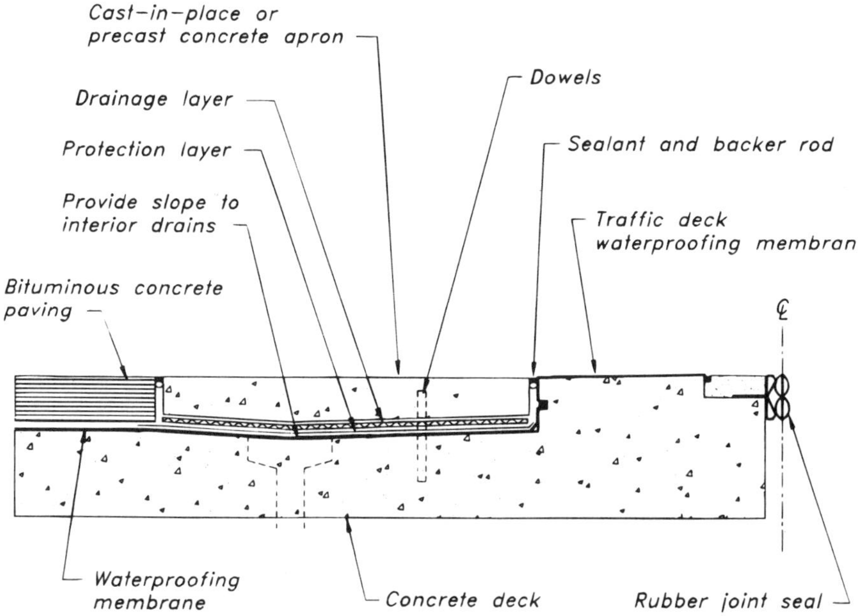

FIG. 9—*Surface seal design for deck expansion joint that permits maintenance of deck waterproofing termination. Apron stability may be a problem.*

drainage at the membrane level and an overburden and wearing surface that protects the joint seal. Only a slight reduction in performance occurs when a bellows is used as the seal, provided the waterproofing system, drainage, overburden, and wearing surface are appropriate. The potential for less performance is due to the lower reliability of the transverse joints in the bellows. The bellows must have factory fabricated corners and transitions in order to provide a reliable seal; particular care in specifying these fabricated seals is necessary in order to get them on the job. Field formed sealant, pre-formed sealants, compression gaskets, and gaskets in metal flanges do not provide protection against leakage appropriate for occupied buildings; these seals are most appropriate for parking garages and require significant routine maintenance.

Certain concrete/elastomer gasket systems are developing satisfactory performance records in parking garage plaza applications. When used with deck coatings, all the criteria for a seal can be satisfied. When used as a curb seal with waterproofing below bituminous paving, waterproofing continuity and the requirement for maintenance are marginally satisfied.

References

[1] *Sweet's Catalog File*, 1965 through 1992.

[2] American Concrete Institute, *ACI 209, Prediction of Creep, Shrinkage, and Temperature Effects in Concrete Structures*, 1986.

Michael A. Lacasse,[1] *Lesia M. Beznaczuk,*[2] *and Krishnan Gowri*[3]

Sealant Selection Using an Expert System

REFERENCE: Lacasse, M. A., Beznaczuk, L. M., and Gowri, K., "**Sealant Selection Using an Expert System,**" *Science and Technology of Building Seals, Sealants, Glazing, and Waterproofing: 3rd Volume, ASTM STP 1254,* James C. Myers, Ed., American Society for Testing and Materials, Philadelphia, PA 1994, pp. 71-81.

ABSTRACT: Sealants used on the exterior of the building envelope must fulfill their performance requirements in a very demanding location within the enclosure system, namely joints. For this reason they tend to be prone to premature failure if not properly matched to the surrounding conditions and materials. Sealant selection is a complex and time-consuming process involving the simultaneous consideration of a multitude of performance criteria. A systematic approach that enables a more comprehensive consideration of the various criteria is therefore important. The prototype expert systems are an attempt to do this. In this paper, a review is made of the main features of the initial prototype sealant selection expert system, including a brief discussion of the decision making process, the assumptions and inherent limitations of the system. Of particular interest in this paper is a description of the advanced prototype system being developed which provides significant improvements over the initial system. This advanced system will include an extensive database from which the following information can be extracted: weather data, sealant performance data, and graphic details database on sealant application.

KEYWORDS: sealant, sealant selection, expert system, building joint design

Sealants that are used on the outside of the building envelope must fulfill their performance requirements in one of most demanding locations within the enclosure system, namely joints. For this reason they are prone to premature failure if not properly matched to the surrounding conditions and materials. Sealant selection is a complex and time-consuming process involving the simultaneous consideration of a multitude of performance criteria. More often than not, one or more factor is likely to be left out in the selection process. A systematic approach that enables a more comprehensive and ready consideration of the various criteria is therefore important.

Several guides for sealant selection intending to provide such a systematic approach exist [*1–5*]. These efforts have been produced by sealant manufacturers, related technical associations, government bodies, as well as researchers and experts in the field. An early effort was made to review and combine this material to produce a comprehensive guide to sealant selection [*6*]. The development of this manual helped establish that the choice of an appropriate sealant, from the point of view of the architect, engineer, or specifier, needed to be based not only on a consideration of potential movement at the joint, but also, on other equally important factors known to affect sealant long-term durability.

[1] Research Associate, Building Science Engineer, National Research Council Canada, Institute for Research in Construction, Materials Laboratory, Ottawa, Ontario, Canada K1A OR6.

[2] Senior Building Science Engineer, Trow Consulting Engineers Ltd., 6 Antares Drive, Unit 7, Ottawa, Ontario, Canada, K2E 8A9.

[3] Associate Professor, Centre for Building Studies, Concordia University, 1455 de Maisonneuve Blvd. W., Montreal, Quebec, Canada, H3G 1M8.

These early efforts demonstrate that no adequate methods exist, at least not by standard test methods, to qualify the life expectancy of sealants based on such factors as, resistance to extension-compression cycling or resistance to ultraviolet radiation. Furthermore, the interrelationship of the two variables mentioned has not yet been made clear, making it difficult to establish the life expectancy of various products and hence the consequences of a particular choice of sealant.

Choosing a sealant is not a trivial matter in light of the fact that despite the low cost of installing sealants in relation to the overall cost of a building, the cost of sealant replacement and repair of damage may be considerably more than that of the installation costs. It is evident that a knowledge of the importance of these factors to the durability of a sealant is of prime importance, however, the only means that exists at this time to obtain such information, is through consultation with an expert practitioner in the domain. That is to say, presently an informed decision is based on information acquired through a practice in the trade.

Knowledge Based Expert Systems

Knowledge based expert systems (KBES) are particularly well suited to the incorporation of heuristic knowledge and this is one of the characteristics that distinguishes it from a conventional (algorithmic) computer program [*7–10*]. The use and development of KBES in engineering has increased steadily in the past fifteen years because they provide advantages in solving problems that require cost effective analysis in a timely manner [*7*]. Furthermore, they can readily accommodate the type of heuristic information in the form of rules of thumb, assumptions, simplifications based on experience, and related expert information that can not be processed by conventional means [*8*]. To some extent, they emulate the decision making process of an expert in the domain.

Hence, KBES are interactive computer programs that contain heuristic knowledge, in the form of rules, as well as factual knowledge in a particular domain (that is, domain knowledge) and perform tasks normally done by an expert [*7,8*]. A schematic of the computer "architecture" that typifies KBES is shown in Fig. 1. The focus in KBES is on knowledge rather than data, as in conventional programs, and emphasis is placed on processing information ("domain" knowledge) by an inference mechanism ("control" knowledge) that controls the manner in which the domain knowledge is applied. Thus "domain" knowledge and "control" knowledge are implemented separately allowing various facilities (that is, explanation and knowledge acquisition facilities) to display, search or modify the domain knowledge base without affecting the manner in which this knowledge is processed by control knowledge (that is, inference "engine"). The control knowledge is also used to modify and expand the context, in which the context contains information about the problem currently being solved. Another feature of KBES programs is that there is some explanation facility to convey to the user the inference actually used in arriving at a solution to a problem. This makes such type of systems particularly useful in the training of non-experts in an experts particular domain.

Expert systems have been used in a variety of engineering applications [*7,9,10*]. For example, in civil engineering, problem solving programs have been developed in the areas of structural analysis and design, code checking, as well as in the areas of geotechnical and transportation engineering and construction management [*7,9*]. More recently, emphasis has been placed on the development of systems for the inspection, repair and maintenance of various materials or structures [*10*] and expert systems are increasingly being used in a number of related fields as was demonstrated at a recent congress of the CIB [*11*].

In this paper, a review is made of the main features of the initial prototype sealant selection expert system [*12,13*], including a brief discussion of the system architecture, the decision making process, and the assumptions and inherent limitations of the system. An example is

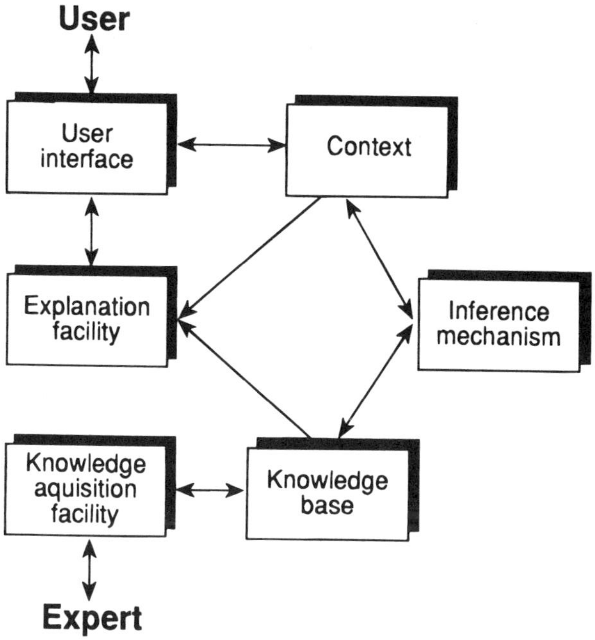

FIG. 1—*Computer architecture of an expert system* [7].

provided to demonstrate the potential usefulness of the prototype system and to indicate the input and output formats used in this program. Of particular interest in this review is a description of the advance prototype system that provides notable improvements over the initial system. This advanced system will include an extensive database from which the following information can be extracted: weather data, sealant performance data, and a graphic details database on sealant application.

Initial Prototype System

The initial prototype expert system was developed from the Sealant Selection Manual [6] as a rudimentary tool for simplifying the selection of a sealant specifically for the purpose of sealing joints on the exterior walls of buildings. It assists in the selection of one or more of the following nine sealants: butyl, skinning; butyl, non-skinning; hypalon; acrylic; polysulphide, 2-part; polysulphide, 1-part; polyurethane, 2-part; polyurethane, 1-part; silicone.

Through a user friendly interface, the system uses a series of questions to prompt the user to provide contextual information, such as input design conditions and sealant performance criteria. Figure 2 shows an example of user input for specifying the surface color of cladding. A list of choices is provided to the user along with a brief description of the performance parameter, however, a separate explanation ("help") facility is not provided. The performance criteria included in the prototype system are: life expectancy, exterior service temperature limits, cladding material, cladding surface color, cladding heat capacity, joint movement, cladding coefficient of thermal expansion, sealant resilience, period or frequency of extension/compression cycling, sealant color range, sealant surface dirt pickup, puncture and tear resistance, surface preparation, surface primer, and tack-free time.

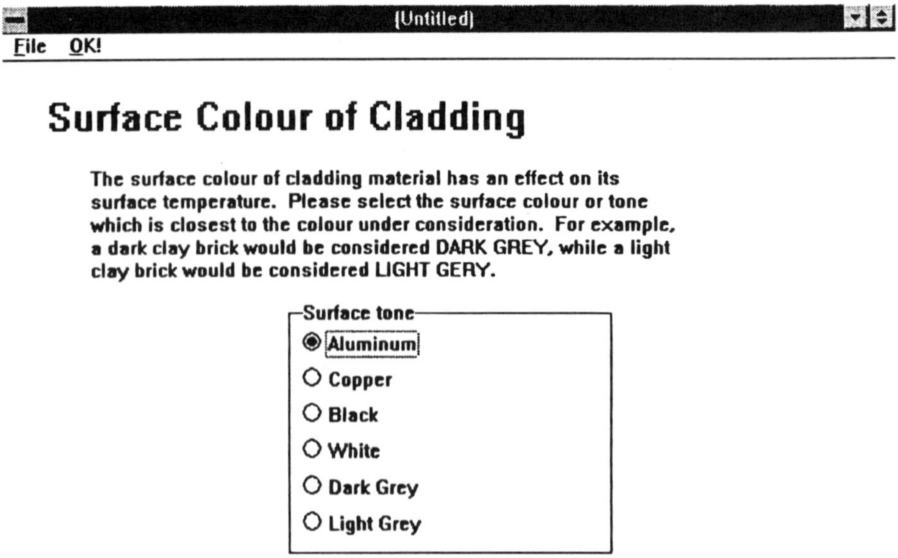

FIG. 2—*Example of user input for the initial prototype system.*

Several parameters do not have to be input by the user, as the knowledge base calculates the values from other related criteria given by the user. These calculated parameters are: maximum and minimum surface temperatures, effective panel lengths, suggested joint widths, calculated and maximum joint movement, and UV resistance required. The values of these parameters are summarized in an output form named "intermediate values." An example is shown in Fig. 3.

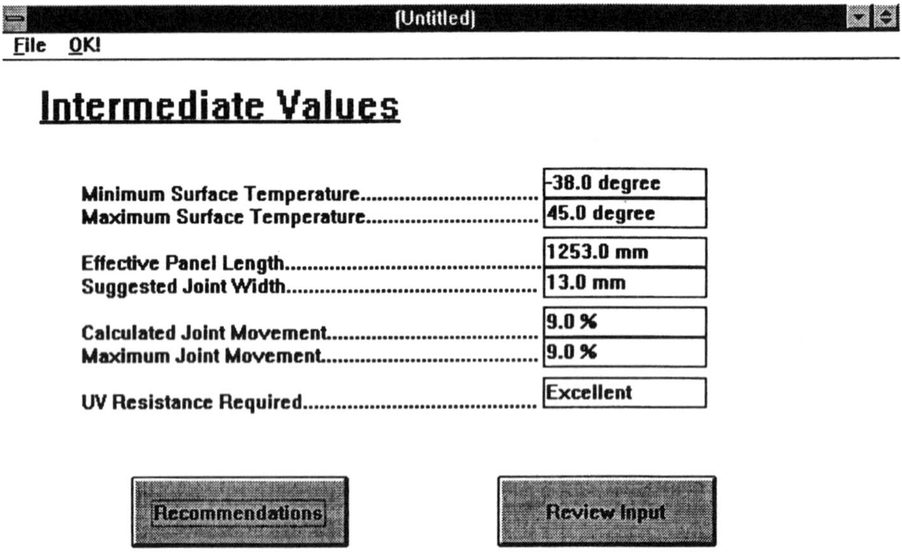

FIG. 3—*Example screen for intermediate values.*

The inference mechanism used in this program determines the relative performance of any of the nine sealants within the system. It assigns a rating (performance factor—PF) for each category of criteria ranging from 0 to 1. This rating is related to the degree to which the sealant performs for a given criteria (e.g., low performance, PF = 0.1; high performance, PF = 0.9 to 1.0). In the selection process, all sealants are initially assigned a value of 100 and thereafter this value is reduced in relation to the sealant's capacity to meet specific performance criteria. In other words, initial values are multiplied by the various factors, and once the system passes through all the required criteria, the cumulative total of all the ratings for each sealant is calculated by the system and presented to the user. This is essentially a process of elimination, as sealants which do not achieve the user's required level of performance are either eliminated or are assigned a relatively low performance factor.

Based on this process of elimination, the system classifies the information and presents the user with a list of recommendations on sealant selection, as shown in Fig. 4. Aspects related to relative costs are kept separate such that a selection can be made based solely on performance characteristics, or based on a combination of both performance characteristics as well as the relative cost of the sealant. In order to assist the user in reviewing the input data, an input summary screen, as shown in Fig. 5, is provided.

Programming Assumptions

Certain assumptions and simplifications were made to limit the amount of input data required from the user, which were:

i. The environmental conditions around the building under consideration are assumed to be uniform across the entire surface of the building envelope.

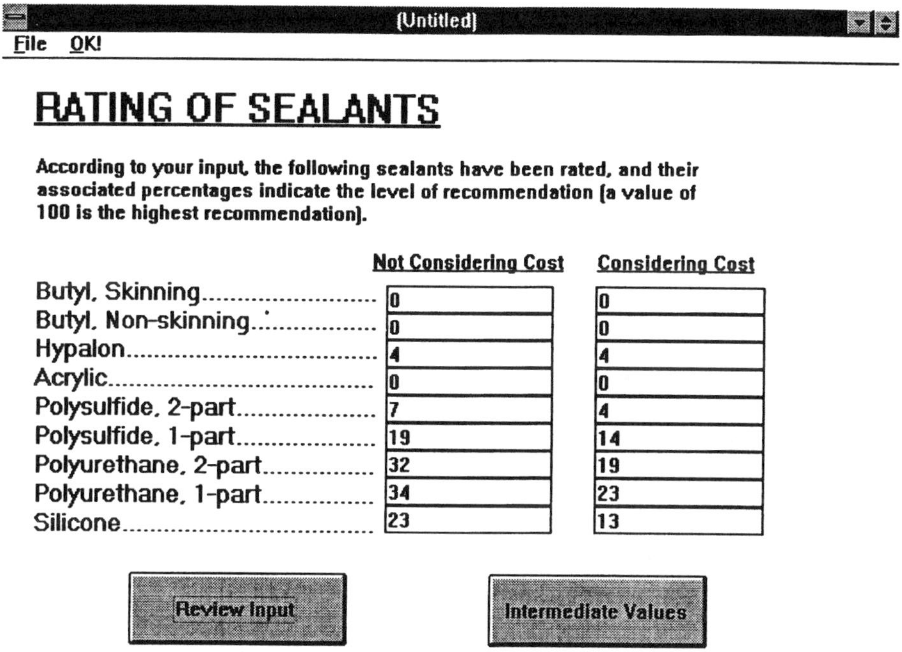

FIG. 4—*Example screen for recommendations.*

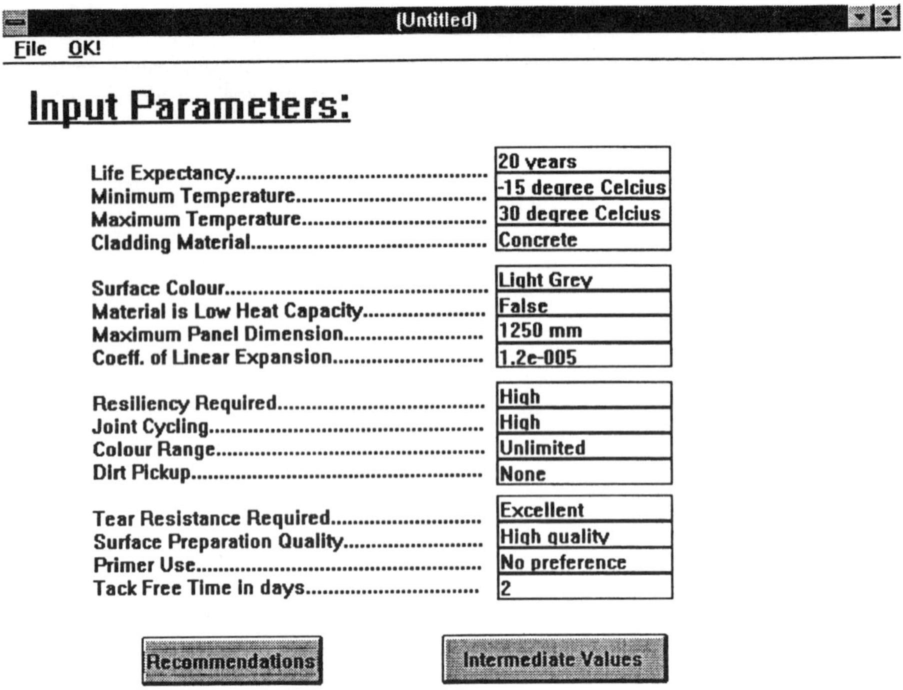

FIG. 5—*Example screen for input parameters.*

ii. Cladding differential movement due to moisture absorption or drying is disregarded as well as any effects due to structural frame movement brought about by loading conditions.
iii. Minimum joint widths are suggested according to panel sizes, following the general rule that the larger the cladding panel the wider the joint should be.
iv. The required ultraviolet (UV) resistance depends only on its required service life. The required UV resistance is assigned, within the program, to service life expectancy accordingly.
v. The coefficient of thermal expansion for four typical cladding materials (masonry, concrete, aluminum, and copper) is given specific values by the system.
vi. Certain performance factors are more critical to overall sealant performance and service life than others. In this respect, less critical performance factors were assigned higher ratings (that is, 0.8 to 1.0) and more critical factors were assigned the entire range of ratings (that is, 0.1 to 1.0).

Program Limitations

Although this initial prototype expert system is quite comprehensive in terms of the performance criteria and design aspects it covers, there are a few limitations to the system in its present form:

i. The system is limited to making recommendations with respect to the nine generic based sealants only.

ii. Recommendations are made for an appropriate sealant choice in a given environment on the basis of the rating system. The performance factors for each sealant are specific to the present program and cannot, as yet, be altered during a consultation session.
iii. The system does not accommodate for joint widths other than the minimum ones recommended by the system (13 mm for cladding panels less than 3000 mm; 16 mm for larger panels). The only accommodation that the system does make is that it recalculates the minimum joint width to limit the maximum joint movement to 25% of the joint width.
iv. The system assumes that a sealant will perform the same when applied to different types of cladding (for example sealants applied to various types of aluminum cladding will perform the same).
v. The system does not allow changing the level of significance of the performance factors.

Advanced Prototype System

The consideration of several performance criteria and ranking of the generic sealants demonstrated by the initial prototype system shows the feasibility and advantages of the expert system approach. However, the initial prototype has a limited scope in terms of the knowledge base, user-interface features and sealant product information. In order to overcome these limitations, an advanced prototype system is currently being developed. The major difference between the advanced and initial prototypes will be the identification and classification of sealant products and performance data in a modular form for easier representation and efficient implementation. A schematic of the new expert system architecture is shown in Fig. 6. The following modules can be identified:

i. User-interface
ii. Inference mechanism (control)

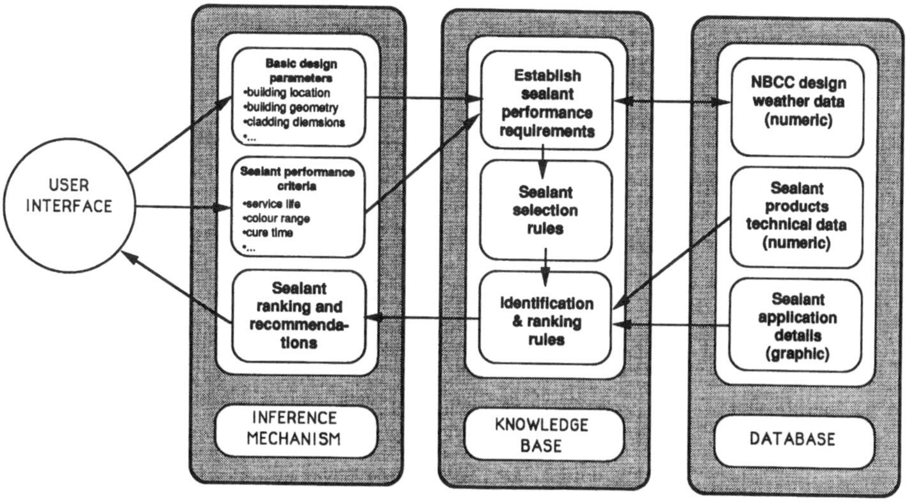

Sealant Selection Expert System - Schematic

FIG. 6—*Proposed expert system architecture for advanced prototype.*

iii. Knowledge-base (domain knowledge)
iv. Database (weather data, sealant properties and graphic details)

User-Interface

The user-interface of the advanced system will consist of simplified input forms, context sensitive help, explanation and reasoning to enable the user to specify preferences and exercise control over the selection process. The user input data, defining the design context, will be minimized to three input forms: (i) building parameters, (ii) desired sealant characteristics and (iii) preferences on sealant performance characteristics. The building parameters would include the location, building type, joint type, cladding material, surface color and length of panels. Many of the sealant selection criteria can be established knowing the building parameters and deriving the appropriate information from the knowledge base and database. For example, the location of a building can be used to obtain the minimum design temperatures, degree days and relative humidity. Similarly, the joint type and cladding material data can be used to select certain generic sealant categories based on the movement capability of the sealant. The desired sealant characteristics would include life expectancy, requirements for priming, tack free time, puncture and tear resistance, etc. These are essentially a combination of quantitative and qualitative criteria which will be used as a basis to search the sealant database for appropriate products Finally, the user can specify preference on performance characteristics such as the cost and color range.

Inference Mechanism

The control knowledge of the proposed system attempts to interface the user-input information to the knowledge base and database. A complete description of the design context is established by the inference mechanism by determining the nature of characteristics and deriving their values appropriately. For example the calculation of surface temperatures would involve obtaining user input of building location, retrieving design temperatures from the database, and accessing the knowledge base for calculation procedures. Once all the values are established, the inference mechanism would attempt to match these values with a database template to identify appropriate sealant products. Finally, sealant products will be ranked based on user specified preferences. The inference mechanism will be implemented to follow a systematic approach which can be reasoned and explained to the user.

Knowledge-base

The knowledge base provides rules-of-thumb and calculation procedures to establish quantitative performance requirements for sealant selection. For example, material compatibility information such as adhesion characteristics will be used to eliminate or select a particular generic category of sealants. In addition, the knowledge base would contain procedures to calculate the surface temperatures, and maximum joint movement. The knowledge base information will be extracted from interviews with experts, design manuals [*14–16*] and other research literature [*17,18*].

Database

The database of the proposed system consists of three major sub-components. The first is a weather database containing design temperatures, degree days, relative humidity and possibly of weather information relevant to assessing the performance of sealants. For the most part, this data may be obtained from the supplement to the National Building Code of Canada [*19*].

The second sub-component is the sealant product performance data which can be classified as quantitative, qualitative, or descriptive in nature. Some of this data is readily available from sealant manufacturers, whereas other qualitative and descriptive information may be obtained from experts and design manuals. The sealant products database will be organized in two levels representing generic and specific products. The first level will identify the generic sealant categories and characteristics. The second level will consist of specific sealant products with their associated technical performance data. Figure 7 shows a typical database structure of the proposed implementation. The quantitative performance data includes sealant costs as well as relevant sealant technical properties such as movement capability, installation and surface temperature ranges and curing time. These are numerical values, which, in certain cases, are based on standard test methods and are often provided by manufacturers. The qualitative performance characteristics include the names of compatible materials, available colors and surface preparation quality. Finally, descriptive performance characteristics are provided for explanation and reasoning purposes. This knowledge will be used to justify the selection or elimination of specific categories or products.

The final sub-component of the database is graphic details of sealant installation. It is often critical to provide the sealant installation sequences and construction details. Such information will be available for each sealant product and other general guidelines will be provided as well. In the advanced prototype, it is proposed to investigate the feasibility of incorporating such graphic information as part of sealant selection recommendations which the user can either view or print to assist in specification writing.

In order to implement the advanced prototype, many expert system development tools available in the personal computer environment have been reviewed. Level 5 Object, operating

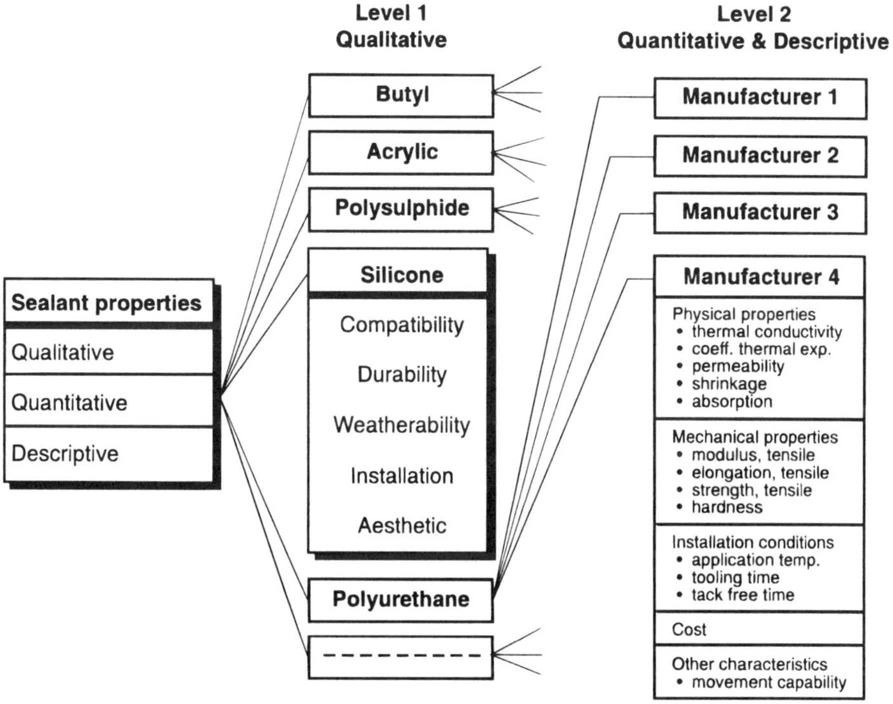

FIG. 7—*Proposed database structure for advanced prototype.*

under the MS-Windows [*20*] environment, has been chosen as the best suited to implement the proposed system. User-input screens, database structure and knowledge acquisition are presently in the process of being completed. Field testing and validation of the system will then be undertaken.

Summary

Expert systems for assisting in the selection of sealants have been developed with the purpose of providing a comprehensive review of various performance criteria such that the task of choosing an appropriate sealant is both less time consuming and is based on a knowledge of factors influencing the choice of a particular product. The development of the initial prototype system was useful in demonstrating the feasibility and advantages of using the expert system approach to perform this often complex task. However, the initial system has a limited scope in terms of knowledge base, user interface features and sealant product information. The advanced prototype system, currently under development, affords significant improvements to the initial system and will offer a comprehensive treatment of joint design and sealant selection based on information obtained from current specification guides, published expert advice, as well as research literature. The proposed advanced system will provide an extended database from which weather data, sealant performance data and graphical details of typical sealant application and installation methods can be obtained.

References

[*1*] Handegord, G. O., Karpati, K. K., "Joint Design and Sealant Selection," *Canadian Building Digest*, No. 155, National Research Council Canada, Ottawa, 1973, p. 4.

[*2*] Koppes, W. P., "Curtain Walls - Sealants," *Time Saver Standards for Architectural Design Data*, 5th ed., J. H. Callendar, Ed., McGraw-Hill, N.Y., 1974, pp. 435–466.

[*3*] Baldwin, B., "Selecting High-Performance Building Sealants," Plant Engineering, 1976, pp. 56–64.

[*4*] *Guide to the Selection of Constructional Sealants*, BS6213:1982, British Standards Institution, 1982, p. 28.

[*5*] *Sealants: The Professional's Guide*, Sealant and Waterproofers Institute, 1984, p. 67.

[*6*] Beznaczuk, L. M. and Lacasse, M. A., *Building Envelope Joint Design and Sealant Selection Manual*, Centre for Building Studies, 1988, unpublished manuscript.

[*7*] Maher, M. L., Ed., *Expert Systems for Civil Engineers: Technology and Application*, ASCE Technical Council on Computer Practices, Committee on Expert Systems, 1987.

[*8*] Fenves, S. J., "What is an Expert System," *Expert Systems in Civil Engineering:* Proceedings of a symposium sponsored by the Technical Council on Computer Practices of the ASCE and the ASCE National Convention in Seattle, Washington, April 1986, C.N. Kostem and M.L. Maher, Eds., ASCE, N.Y., 1986, pp. 1–6.

[*9*] Adeli, H., Ed., *Expert Systems in Construction and Structural Engineering*, Chapman and Hall, London, 1988, p. 330.

[*10*] Fenves, S. J., "Expert Systems; Expectations versus Realities," *Expert Systems in Civil Engineering, International*, Association for Bridge and Structural Engineering (IABSE) Colloquium, Bergamo, Italy, October 1989, IABSE, Zurich, Report Vol. 58, 1989, pp. 1–19.

[*11*] World Building Congress - CIB '92, Montreal, Canada, May 1992, International, Council for Building Research Studies and Documentation (CIB), Poster Book Vol. 2, Section SB, p. 498.

[*12*] Beznaczuk, L. M. and Lacasse, M. A., "Prototype Sealant Selection System is an Expert Specifier," Canadian Consulting Engineer, Mar./Apr. 1992. pp. 30, 32–33.

[*13*] Beznaczuk, L. M. and Lacasse, M. A. , "Building Envelope Sealant Selection: An Expert System," Construction Canada, Vol. 34, No. 5, 1992, pp. 37–39.

[*14*] *Manual of Good Practice in Sealant Application*, CIRIA Special Publication 80, Construction Industry Research and Information Association, British Adhesives and Sealants Association, London, U.K., 1991, p. 58.

[*15*] Beall, C., Cavalier, J. J., Gardiner, B., "Joint Sealers," CSI Monograph Series, Construction Specifiers Institute, Virginia, 1991, p. 51.

[*16*] Nicastro, D. H., "Parameters for Comparing High-Performance Sealants," Construction Specifier, 1990 pp. 54–61.

[*17*] Amstock, J. S., "*Sealants*," *Kirk-Othmer Encyclopedia of Chemical Technology*, J. I. Korschwitz and M. Howe-Grant, Eds., Wiley, N.Y., N.Y., 1992, pp. 354–377.
[*18*] Beech, J. C., "*The Selection and Performance of Sealants*," Building Research Establishment Information Paper IP25/81, Building Research Station, Garston, U.K., 1981, p. 5.
[*19*] National Building Code of Canada 1990, 10th ed., NRCC No. 30619, National Research Council Canada, Ottawa, 1990, p. 423.
[*20*] Microsoft® Windows™.

Testing

Imad L. Al-Qadi[1] *and Saad A. Abo-Qudais*[1]

Test Method for Evaluating Pavement Sealants Under Simultaneous Cyclic Shear and Normal Deflections

REFERENCE: Al-Qadi, I. L. and Abo-Qudais, S. A., "**Test Method for Evaluating Pavement Sealants Under Simultaneous Cyclic Shear and Normal Deflections**," *Science and Technology of Building Seals, Sealants, Glazing, and Waterproofing: 3rd Volume, ASTM STP 1254*, James C. Myers, Ed., American Society for Testing and Materials, Philadelphia, 1994, pp. 85–94.

ABSTRACT: A new method to evaluate the performance of sealants, used in rigid pavement joints, was developed. A special fixture was designed to transfer cyclic in-line deflection, applied by a testing machine, to simultaneous cyclic normal and shear deflections, at a specific ratio, on a sealant sandwiched between two-51 mm portland cement mortar cubes. The developed fixture was used to evaluate the performance of three commercially available one-component sealant types (A, B, and C): A is a low modulus silicone, B is a self leveling silicone, while C is a polyurethane used with primer.

The sealants were evaluated under both cyclic compression/shear and tension/shear tests at 25 Hz. Each sealant was evaluated at two joint widths, 6.5 mm and 19 mm. Analysis of test results indicated that sealant C performed the best with no failure after 500,000 cycles in all tests, while the performance of sealant B was the worst with cohesive failure in most cases. Sealant A, on the other hand, failed adhesively at a joint width of 6.5 mm. All sealants, not surprisingly, performed better in cyclic compression/shear than in cyclic tension/shear.

KEYWORDS: sealant, pavement joint, compression/shear, tension/shear, normal deflection, horizontal deflection, cyclic deformation

The low tensile strength of Portland cement concrete makes it necessary to provide joints for many concrete structures including rigid pavements. Rigid pavement joints may be classified into three categories: transverse joints (expansion and contraction joints), longitudinal joints (center line and edge joints), and construction joints. Many material types have been used to seal the joints, including tar, asphalt, rubber, inorganic-elastomeric, polysulfide, silicone, polyurethane, rubberized asphalt, and preformed compression seals [*1*].

Failure in joint systems normally causes service life reduction of rigid pavements. A high percentage of deterioration in concrete pavements occurs at or near the transverse contraction joints due to inadequately sealed or failed joints [*2*]. The joints may fail if one of the following problems in the sealant occurs: adhesive failure, cohesive failure, extrusion or intrusion failure, and/or impregnation of incompressible materials.

Joint deterioration may occur in several ways: for instance, in cold regions, joint spalling is usually caused by infiltration of water and incompressible materials into the joints. The incompressible foreign materials build up in the joints and when the temperature rises, the joints close. Thus, tremendous pressures build up and the weakest joint in a series offers stress

[1] Associate Professor and Graduate Research Assistant, respectively, The Charles E. Via Department of Civil Engineering, Virginia Polytechnic Institute and State University, Blacksburg, VA 24061-0105.

relief to the compression buildup in the pavement by shattering or by buckling upward; this may occur gradually or suddenly cause blowups and joint spalling [3]. This causes pavement roughness, which may lead to severe damage, such as slab crushing and/or cracking.

Pumping is another problem that may be caused in part by the intrusion of water through failed joints. Pumping occurs when infiltrated water carries fine particles from the pavement foundation and/or shoulders and ejects them onto the pavement surface during traffic loading. Pumping becomes a serious problem when a large volume of material is displaced leaving unsupported slab areas and thus pavement fault occurs [4].

The use of deicing salts in the United States also causes serious damage to pavement joints. Chlorides in deicing salts penetrate failed joint sealants and corrode the reinforcing dowels and/or tie bars. The expansive steel corrosion products (which has a volume of almost six times the original product) results in increased internal stresses that ultimately crack the joint concrete.

Shear movements, resulting mainly from traffic and occasionally from temperature variations, contribute to sealant failure, especially if the sealant initially does not possess sufficient elastomeric properties or if it has lost its elastomeric properties through aging. The degree of loss of the elastomeric properties of the joint sealants depends mainly on temperature, weathering, joint width, and load repetition.

A series of investigations were performed to evaluate many sealant types. The laboratory tests conducted on sealants have concentrated on tensile, compression, bond, penetration, flow, stress relaxation, shear fatigue, and solubility. For example, Collins et al. [1] evaluated five different sealants, two-part polysulfide, silicone, hot poured rubberized asphalt, polyurethane, and preformed sealant. Silicone performed better in elongation while rubberized asphalt performed better in flow, stress relaxation, and penetration. However, low modulus silicone performed the best in accelerated aging testing [5]. Environmental effects were also studied by Jones et al. [6] and correlated to strength loss while the effects of joint width on sealant deformability were investigated by Kuenning [7] at different temperatures.

Field observations have also been used to assess the service life of joint sealants. For example, it has been reported that the service life of sealants in Louisiana is from zero to 76 months; most failures being adhesive [8]. Collins et al. [1] reported that Dow Corning 888 silicone sealant had excellent performance in the field, while rubberized asphalt was rated very poor. Dow Corning 888 silicone and Crafco Roadsaver 231 and 221 sealants performed exceptionally well over two years, while rubberized asphalt performed poorly and showed adhesive, embedment, and extrusive failures. Rubberized ashpahlt performance was also poor in a study conducted over 15 years in Michigan; neoprene performed better than rubberized asphalt in that study [9]. Wolters [3] reported poor performance of preformed sealants compared to liquid sealants. Other studies have analyzed collected data from questionnaires and interviews with experts [4].

At present, pavement engineers have to choose from a large number of sealants. While extensive research has been conducted on the use of joint sealants in concrete pavements, acceptable sealing of joints continues to be one of the most debated issues in concrete pavement design and maintenance [7].

An examination of available literature shows that sealant acceptance specifications requires laboratory testing which ignores the effect of shear deflection due to vehicular loading. This may explain the poor field performance of some sealants that had previously met the laboratory testing requirements. It is well recognized that joint wall-sealant failures are caused by repeatable normal pavement joint opening and closing due to slab expansion and contraction caused by variation in temperature. However, sealant failure resulting from cyclic vertical shear in the joint caused by heavy truck trafficking has not been addressed. Currently, there is no standard method to evaluate the performance of sealant in a manner that simulates field loading conditions.

The objective of the study reported in this paper was to develop a new testing technique to evaluate joint sealant performance. The technique was used to evaluate the potential field performance of three types of joint sealants. The sealants were exposed to simultaneous cyclic shear and normal deflections at different temperatures. As the joint width may have an influence on joint performance, two different joint widths were considered for evaluation. The study, reported in this paper, is part of an ongoing investigation which is also considering other testing fixture design and environmental effects such as freezing and thawing and deicing salt applications on the service life performance of joint sealants.

Experimental Program

Fixture Design

In an attempt to simulate field deflections of joint sealants, normal deflections due to temperature variations and shear deflections due to wheel loads, a special fixture was designed; a schematic diagram is presented in Fig. 1. The test specimen consists of a sealant sandwiched between two 51 mm portland cement mortar cubes, held by the test fixture at different angles.

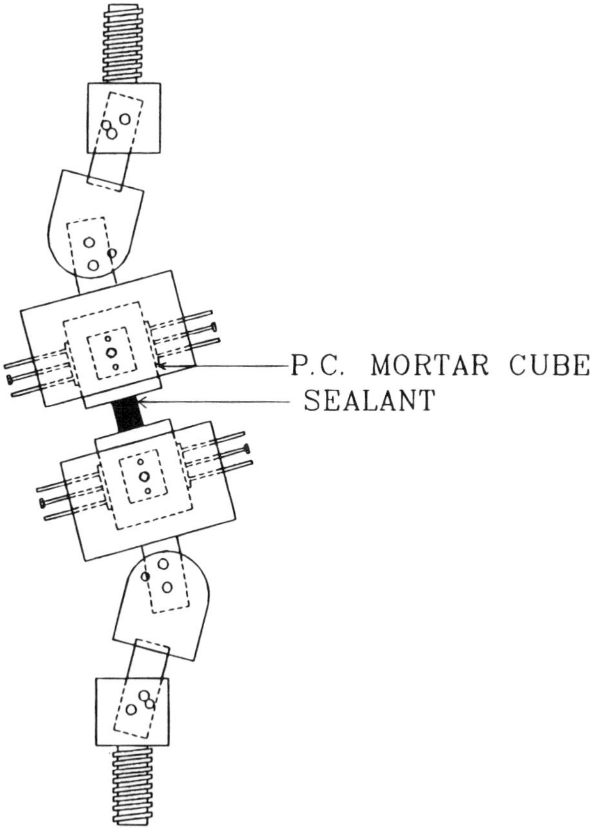

FIG. 1—*Schematic diagram of designed fixture.*

By holding the specimen at a specific angle, the in-line deflection applied by the testing machine is transferred to applied normal and shear deflections on the sealant at a ratio corresponding to the angle at which the specimen is held. The fixture can be adjusted to different ratios of shear to normal deflections. The two cubes of the specimen are held inside two steel boxes by adjustable plates which secure the cubes and prevent specimen rotation.

Tested Sealants

Three types of rigid pavement sealants were evaluated in this study: sealants A, B and C. Sealant A is a one part, cold applied low modulus silicone that readily extrudes over a wide range of temperature. When it cures, it is thought to produce a durable, flexible, low-modulus sealant. This sealant is easy to use with gunability at a wide temperature range. It also bonds to concrete without using a primer and is thought to have good weathering resistance characteristics.

Sealant B is a one-component, cold-applied, self-leveling silicone that also readily extrudes over a wide temperature range. The main expected features of this sealant are ease of application, good weatherability, and bondability to concrete without using a primer. Sealant C is a one-part cold applied polyurethane that bonds to concrete by means of a primer to be applied to the joint concrete surface one hour before joint sealing.

Specimen Preparation

Each test specimen consists of a sealant (joint dimensions: 19 × 19 × 51 mm or 6.5 × 6.5 × 51 mm) sandwiched between two 51 mm portland cement mortar cubes prepared in accordance with ASTM C 109 and moist cured for 28 days. The water to cement (w/c) ratio of the portland cement mortar was 0.49 and its average 28-day compressive strength was 30 MPa with a standard deviation of 0.5 MPa. The sealant shape was formed by placing two removable styrofoam pieces between the two mortar cubes, forming a fixed reservoir volume section between them (joint dimension), and held in place with duct tape wrappings. The sealant was then gunned between the two styrofoam pieces through a cut made in the duct tape. The duct tape and the styrofoam pieces were removed after five days, and the specimen was allowed to cure for an additional 25 days.

A testing program was developed to validate the testing fixture design. To accomplish this task, three types of sealants were evaluated, each at two different joint widths: 19 mm and 6.5 mm. Specimens with 19 mm joint width simulated, in the field, an expansion and contraction joint between 18.3 m reinforced concrete slabs. Specimens with 6.5 mm joint width simulated a joint in plain (unreinforced) concrete pavement. The normal slab length for plain concrete pavement is 4.6 m.

Testing Program

In an attempt to simulate field loading conditions, the specimens were tested under simultaneous cyclic normal and shear loading (deflection controlled). An Instron machine model 1331 was used to apply the deflection to the specimen through the test fixture. The test fixture transfers the in-line deflection applied by the Instron machine into normal and shear deflections on the joint sealant. For the 19 mm joint width, a normal deflection of 4.2 mm was used, which corresponds to the maximum expansion and contraction of a 18.3 m slab length due to a 28°C temperature change. Although shear deflection in the field is highly dependent on subgrade reaction and/or subbase strength, the shear deflection for testing the 19 mm joint width was selected to be 1.4 mm to represent the maximum deflection caused by truck trafficking. The normal to shear deflection ratio was controlled by the test fixture angle, 18.6

degrees. It has to be pointed here that shear deflections in the field, caused by vehicular loading are not pure shear especially if concrete slabs are exposed to warping and/or curling stresses which are daily and seasonal reactions to temperature and moisture variations between the top and bottom of the concrete slabs.

A preliminary study was performed to choose a proper deflection frequency for use during testing. This preliminary study was performed on sealant A with a 19 mm joint width. Twelve specimens were tested either in compression or tension at six different frequencies: 5, 10, 20, 25, 30, and 40 Hz at room temperature. The 25 Hz frequency was selected to be applied throughout the study due to the inaccuracy of load measurements at lower frequencies. This frequency is relatively high compared with traffic loading frequency, however, it would be acceptable to evaluate relative performance of sealants.

During a specific test, the change in load was monitored at a constant normal and shear deflection. Joint sealants were considered to have failed when 20% cohesive and/or adhesive tearing was observed. If no tearing was observed, deflection cycles were applied up to 500 000 cycles.

Because tension on joint sealant occurs during cold weather when slabs contract due to low temperature, tension properties of joint sealant were evaluated at an average temperature of -4 ± 2°C, while compression of joint sealant was evaluated at an average temperature of 40 $\pm$ 2°C. At 40°C, pavement slabs expand and compress sealant in the joint.

Sealant A with a 19 mm joint width was tested in compression/shear; the specimen was heated in an oven at 42°C for two hours; specimen was returned to oven when its temperature decreased to 38°C and reheated to 42°C. The specimen was placed in the fixture, and an in-line compression deflection of 4.4 mm at 25 Hz was applied producing a normal deflection of 4.2 mm and a shear deflection of 1.4 mm. The force required to cause the deflection was monitored at 500 cycle intervals, by an oscilloscope. For tension testing, the specimen was held at -6°C for six hours prior to testing. When the specimen temperature was increased to -2°C during testing, it was re-cooled. The same testing procedure was used for sealant A with a joint width of 6.5 mm using an in-line deflection of 1.5 mm instead of 4.4 mm, normal deflection of 1.4 mm and shear deflection of 0.5 mm.

After every 500 cycles, the sealant was visually inspected and the type and percent of tearing was noted. The same testing sequence was repeated for sealants B and C. All tests were conducted in triplicates.

Results and Discussion

Sealant test results, summarized in Table 1, were plotted as normal or shear stress versus number of cycles. The performance of the three types of sealant investigated under different loading conditions is presented in Figs. 2 through 5, for different joint widths.

For 19 mm joint widths: the stress (both normal and shear) was the smallest for sealant B, while for sealants A and C it was approximately twice as great in the case of compression/shear and almost five times in tension/shear. The test results showed that sealant C, at 19 mm joint width, performed the best in tension/shear and compression/shear as compared to the other two sealants, Figs. 2 and 3.

Figure 2 illustrates that sealant C, when tested in compression/shear at a joint width of 19 mm, resisted 500 000 cycles of a normal deflection of 4.2 mm and a shear deflection of 1.4 mm without tearing. The average stress in sealant C initially was 200 kPa, which decreased to 193 kPa at approximately 1000 cycles and then remained constant during testing up to 500 000 cycles. Sealant A failed adhesively after 15 000 cycles; its initial stress was 205 kPa, which decreased to 204 kPa at 1000 cycles and gradually decreased to 193 kPa until it failed.

TABLE 1—*Test results.*

Sealant Type	Type of Test	Joint Width (mm)	Failed	Type of Failure	No. of Cycles	Normal[a] Stress at Failure (kPa)	Shear[a] Stress at Failure (kPa)
A	Comp	19	Yes	Adhesive	15,000	193.20	64.38
		6.5	No	—	500,000	248.40	82.80
	Tension	19	Yes	Adhesive	7,500	182.85	60.93
		6.5	No	—	500,000	165.60	55.20
B	Comp	19	Yes	Cohesive	1,000	75.90	25.32
		6.5	No	—	500,000	41.40	13.80
	Tension	19	Yes	Cohesive	1,000	41.40	13.80
		6.5	Yes	Cohesive	100,000	34.50	11.52
C	Comp	19	No	—	500,000	193.20	64.38
		6.5	No	—	500,000	345.00	115.02
	Tension	19	No	—	500,000	220.80	73.62
		6.5	No	—	500,000	420.90	140.28

[a] If no failure was observed, stress is at 500 000 cycles, values are an average of three specimens.

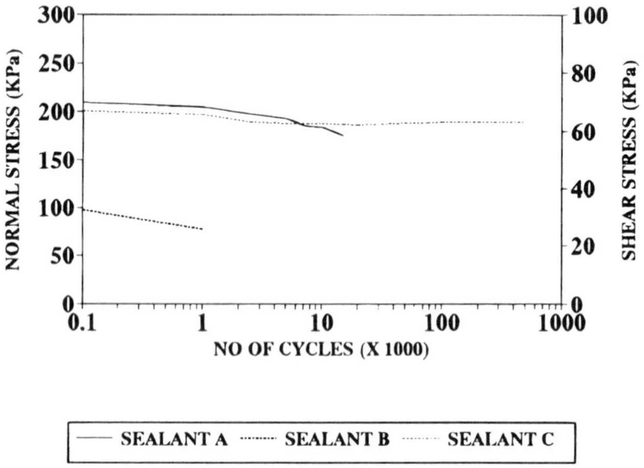

FIG. 2—*Compression/shear stress—number of cycles relationship at 19 mm joint width.*

Sealant B failed cohesively after approximately 1000 cycles ; its initial stress was 97 kPa and decreased to 76 kPa where it failed.

The performance of sealant C was the best. Although all sealants were used as suggested by the manufacturers, sealant C was the only sealant to be used with a primer. The adhesive failure of sealant A after 15 000 cycles was due to the low adhesive bonding strength between the sealant and the portland cement mortar cubes. Note that no primer was used with this sealant. Sealant B, on the other hand, failed cohesively after 1000 cycles at a lower stress than the other two sealants. It was observed that sealant B failed due to its low stiffness, and its low elongation capability, which was suspected to be related to insufficient curing, although all specimens were cured for 30 days before testing. Better curing was noticed when sealant B was exposed to cycles of freezing and thawing [*10*].

The three sealants performed similarly in tension/shear. As shown in Fig. 3, sealant C performed as it did in compression/shear; however, the normal stress was a little greater, 221

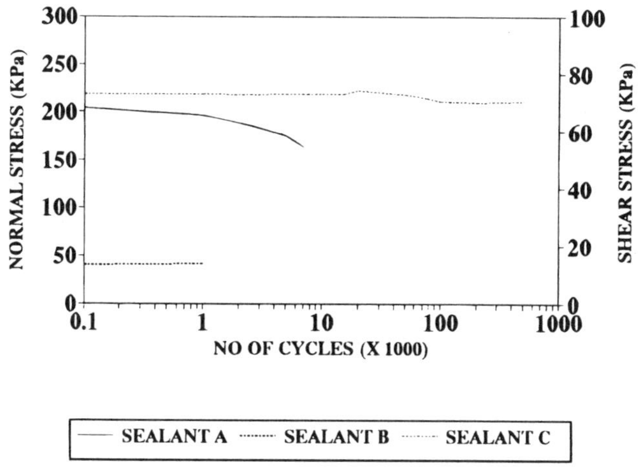

FIG. 3—*Tension/shear stress—number of cycles relationship at 19 mm joint width.*

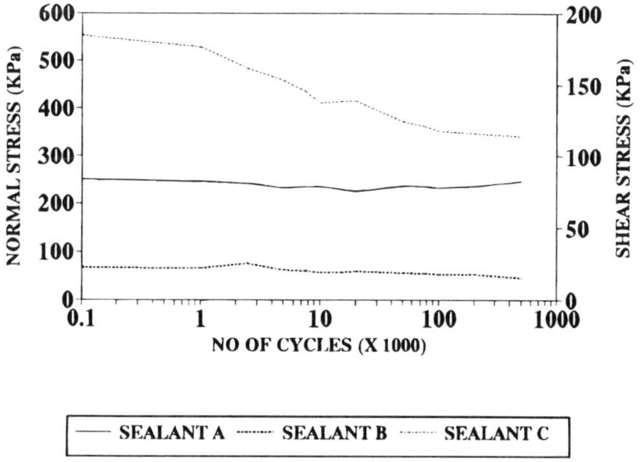

FIG. 4—*Compression/shear stress—number of cycles relationship at 6.5 mm joint width.*

kPa, and remained almost constant throughout the 500 000 cycles. Sealant A exhibited the same stress trend in tension as in compression except it failed at a lower number of cycles, 7500 cycles, at a normal stress of 183 kPa. The performance of sealant B was again poor. The sealant failed at about 1000 cycles, and the normal stress remained constant throughout the 1000 cycles at 41 kPa. The same types of failure noticed in compression occurred in tension. Sealant B failed cohesively, while sealant A failed adhesively.

At 6.5 mm joint width, the test results demonstrated that both sealants A and C performed very well in tension/shear and compression/shear (Figs. 4 and 5). Sealants A and C resisted 500 000 cycles of 1.4 mm normal deflection and 0.5 mm shear deflection without failure. In compression, sealant B resisted 500 000 cycles of deflection, while in tension it failed cohesively after 100 000 cycles. Sealant B, at 6.5 mm joint width, showed better curing than that for 19 mm joint width; however, it was suspected to be not fully cured.

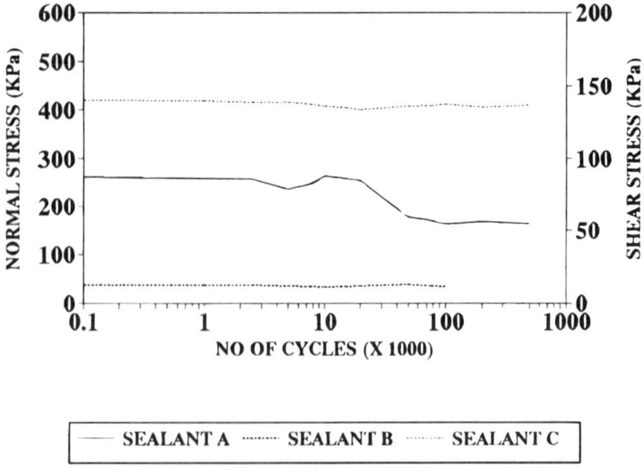

FIG. 5—*Tension/shear stress—number of cycles relationship at 6.5 mm joint width.*

The normal stress was the highest for sealant C, approximately twice that of sealant A, and almost 10 times that of sealant B for both cases: compression/shear and tension/shear. In compression/shear (Fig. 4), the initial normal stress in sealant C was 552 kPa, which decreased gradually to 359 kPa at 500 000 cycles. Sealant A initial normal stress was 248 kPa and remained almost constant for the 500 000 cycles. Sealant B initial normal stress was 69 kPa and decreased gradually to 41 kPa until 500 000 cycles. In tension/shear (Fig. 5), the initial normal stress in sealant C was 421 kPa which remained almost constant for the 500 000 cycles. Sealant A initial normal stress was 262 kPa and remained constant for 25 000 cycles, then fluctuated between 166 and 276 kPa until 500 000 cycles. On the other hand, sealant B had an initial normal stress of 35 kPa and remained constant until 100 000 cycles when it failed cohesively. This performance may be explained by the fact that sealant C was the only sealant which was used with a primer. It was observed during testing that sealant A had low adhesive bonding strength to the joint wall, while sealant B had low stiffness.

In summary, the test results showed that the three evaluated sealants performed in compression/shear better than in tension/shear. Sealant A, at 19 mm joint width, failed in compression/shear at 15 000 cycles, while in tension/shear it failed at 7500 cycles. In both compression/shear and tension/shear tests, at 6.5 mm joint width, sealant A was exposed to 500 000 cycles without failure. For the two evaluated joint widths, 6.5 mm and 19 mm, sealant C was exposed to 500 000 cycles without failing in compression/shear or tension/shear. Sealant B failed at approximately 1000 cycles in both compression/shear and tension/shear for 19 mm joint width, while for 6.5 mm joint width, it failed in tension/shear at 100 000 cycles, and in compression/shear it resisted 500 000 cycles without failure.

The greater resistance of the sealant in compression/shear compared to that in tension/shear is related to the properties of the sealant; sealants, in general, are stiffer and more viscous in compression when dynamically loaded. Also, compression resulted in stronger shear forces at the sealant-joint wall interface than tension.

Summary, Findings, and Conclusions

This investigation is considered the start of a more comprehensive research project. In this study, a new method to evaluate the performance of sealants used in rigid pavement joints was developed. A specially designed fixture was used to apply cyclic normal and shear deflections on a sealant sandwiched between two 51 mm portland cement mortar cubes. The normal and shear deflections were applied simultaneously at a specific ratio controlled by the fixture setup. Under ideal field conditions, shear deflection on sealant is caused by vehicular loads, while the normal deformation (tension or compression) is caused by slab contraction or expansion due to temperature variation. Three sealant types were evaluated using the newly developed setup. A polyurethane sealant applied with a primer (type C) performed, in general, the best followed by a low modulus silicone sealant (type A). However, the performance of a self leveling silicone (type B) was poor and suspected to had not completely cured although it was cured for 30 days before testing. Sealant C required special care in preparing the test specimens due to its low initial viscosity and required up to five days of curing. In general, test results demonstrate the ability of the tested sealants to resist simultaneous applied normal and shear deflections.

Based on the information presented, the following conclusions are made:

1. A new testing technique was developed to evaluate sealant performance under simultaneous cyclic shear and normal deflections.
2. Based on limited test results, sealant C performance ranked the highest, followed by sealant A and then sealant B. Sealant B poor performance was explained by its suspected incomplete cure, although it was cured for 30 days. All sealants were applied as suggested by the manufacturers.
3. Sealant performance in compression/shear was better than that in tension/shear; most failures in tension were adhesive debonding.
4. Sealant performed better at 6.5 mm joint width than at 19 mm joint width. The only sealant failed at 6.5 mm joint width was sealant B in tension after 100 000 cycles compared to failure after 1000 cycles for 19 mm joint width.

Acknowledgment

This research has been partially sponsored by the Center for Adhesive and Sealant Science of Virginia Tech and the Adhesive and Sealant Council, Inc. The authors would like to thank Glen Thomas of the Via Department of Civil Engineering Machine Shop for his assistance in building the fixture. Also, the authors wish to acknowledge the suggestions offered by Alumni Distinguished Professor James P. Wightman of Chemistry Department and Professors Richard E. Weyers and Richard D. Walker of Via Department of Civil Engineering at Virginia Tech.

References

[*1*] Collins, A. M., Magnum, W. D., Fowler, D. W. and Meyer, A. H., *Improvement Methods for Sealing Joints in Portland Cement Concrete Pavements*, Report No. 385-1, Center For Transportation Research, University of Texas at Austin, Austin, TX, 1986, p. 132.

[*2*] Capas, T. L. and Pennock, H. A., *Design, Construction, and Maintenance of PCC Pavement Joints*, No. 19, NCHRP, Washington, D.C., 1973, p. 40.

[*3*] Wolters, R. O., *Field Evaluation of Joint Seal Material*, Minnesota Department of Highway, Report No. ALA(2), 1974, p. 19.

[*4*] Cook, J. P. and Lewis, M. R., *Evaluation of Pavement Joint and Crack Sealing Material and Practices*, Report No. 38, NCHRP, Washington, D.C., 1967, p. 37.

[*5*] Sandberg, L. B., "Comparisons of Silicone and Urethane Sealant Durabilities," *Journal of Materials in Civil Engineering*, ASCE, Vol. 3, No. 4, 1991, pp. 278–290.

[*6*] Jones, G. M., Peterson, D. T. and Vayas, R. K., *Valuation of Preformed Elastomeric Pavement Joint Sealing System and Practices*, Report No. 19, NCHRP, Washington, D.C., 1973, pp. 32–41.

[7] Kuenning, W. H., "Laboratory Tests of Sealers for Sawed Joints," *Highway Research Board*, No. 211, Washington, D.C., 1962, pp. 1–12.
[8] Kinchen, R. W., Temple, W. H., Brut, W. T., Azar, D. G., *Evaluation of Joint Sealant Materials*, Louisiana Highway Research, Louisiana, 1977, p. 66.
[9] Oehlr, L. T. and Bashore, F. J., "Michigan's Experience with Neoprene Compression Seals," *ACI Journal*, No. 70-22, 1973, pp. 214–220.
[10] Abo-Qudais, S. A. and Al-Qadi, I. L., "Performance Evaluation of Sealants Used in Rigid Pavement Joints," Presented at CASS/CCMS Workshop, Virginia Tech, Blacksburg, VA, Oct. 3–5, 1993, p.1.

Pierre Descamps,[1] *Jean Iker,*[1] *and Andreas T. Wolf*[2]

Methods of Predicting the Adhesion of Silicone Sealants to Anodized Aluminum

REFERENCE: Descamps, P., Iker, J., and Wolf, A. T., "**Methods of Predicting the Adhesion of Silicone Sealants to Anodized Aluminum,**" *Science and Technology of Building Seals, Sealants, Glazing, and Waterproofing: 3rd Volume, ASTM STP 1254*, James C. Myers, Ed., American Society for Testing and Materials, Philadelphia, 1994, pp. 95–106.

ABSTRACT: The quality of silicone sealant adhesion to anodized aluminum varies widely. Key parameters that affect adhesion are the degree of surface sealing, the nature of the cleaner, and the time period allowed between cleaning the substrate and sealant application. While the cleaners display different effectiveness in removing organic contamination from the anodized aluminum surface, this effectiveness does not correlate with adhesion quality. It is hypothesized that by adsorbing on the anodized aluminum substrate the cleaning solvents 'condition' the surface for improved silicone sealant adhesion. This conditioning effect diminishes, as the cleaning solvents desorb over time. For a given cleaning solvent, optimum adhesion is observed for a certain degree of surface sealing. Adhesion to unpigmented anodized aluminum substrates can be predicted by measuring the degree of surface sealing according to the ISO 2143 acid etch method. For pigmented anodized aluminum substrates, electrical phase shift measurements can be used to predict adhesion quality.

KEYWORDS: anodized aluminum, surface sealing, cleaning, structural glazing, silicone sealant, adhesion, contact angle hysteresis

Structural glazing is becoming increasingly popular as a means of finishing building facades. In this method, glass, aluminum, stone or other materials are bonded to the supporting curtain wall structure by silicone sealants. This construction technique, initially developed in the United States almost three decades ago [*1*], is now being widely used by architects in North America, Europe, Australia and the Pacific Rim.

Compared with classical lightweight curtain wall construction, structural glazing offers various aesthetic and functional advantages. Bonding the facade panels into place with silicone sealants eliminates the need for exterior mechanical retention members, and therefore allows the design of continuous, free-flowing lines of glass as well as of smooth, uncluttered facades. In the first case, the so-called 'two-sided' method, two parallel edges of the facade panel are adhered to the building frame using a structural silicone sealant. The other two edges are conventionally retained by glazing channels. The two-sided method is the most widely used structural glazing technique. However, there is a trend amongst architects to design structural glazing facades, in which all four edges of a panel are bonded into place without any mechanical retention. In this so-called 'four-sided' method, the structural integrity, and thus the safety of

[1]Technical Service Engineer, and Section Manager, respectively, Sealants Business, at Dow Corning S. A., Parc Industriel, 7180 Seneffe, Belgium.

[2]Global Technology Manager, Sealants Business, at Dow Corning Corporation, Midland, Michigan 48686.

the system, depends on the adhesion of the sealant to the external facade panel and the supporting building frame.

Specially formulated silicone sealants are presently the only materials that possess all necessary properties for structural glazing: reliable adhesion to a variety of building substrates; elastomeric properties that allow accommodation of the thermal and structural movements of the building components; and, finally, adhesive and cohesive properties that are little affected by ultra-violet radiation and other environmental factors.

While the outstanding bonding capacity of structural silicone sealants to glass or glazed surfaces is well documented, the mechanism of adhesion to other substrates is not yet fully understood. Various finishes, applied to the surface of structural curtain wall members, for instance, result in significant changes to the surface properties of the substrate, and, thus, can adversely affect sealant adhesion. Manufacturers of structural silicone sealants therefore test the adhesion to critical substrates on a project-by-project basis.

In Europe, more than 80% of all structural glazing construction is carried out with anodized aluminum frames. Project approval tests carried out within Dow Corning revealed that the quality of silicone sealant adhesion to anodized aluminum can vary widely, depending on the surface properties of the substrate at the time of sealant application, which in turn depend on the initial surface properties, the nature of the solvent used in cleaning the substrate, and the time span between cleaning of the substrate and sealant application.

This paper presents the results of an investigation on how the surface nature of anodized aluminum affects the adhesion of structural silicone sealants. The 'conditioning' effect of cleaning solvents on the anodized aluminum surface is demonstrated and key surface parameter that affect sealant adhesion are identified. Finally, as part of this study, two test methods were developed that allow prediction of the quality of silicone sealant adhesion on anodized aluminum substrates.

Characterization of the Anodized Aluminum Surface

'Anodized' aluminum is manufactured by anodic oxidation of aluminum in a sulphuric acid (H_2SO_4) medium. Its surface constitutes a compact system of hexagonal cells, with each cell containing a pore (Fig. 1). The diameter of the cells formed depends on the process conditions, for example, electrolyte composition, temperature, current density, et cetera, and ranges from

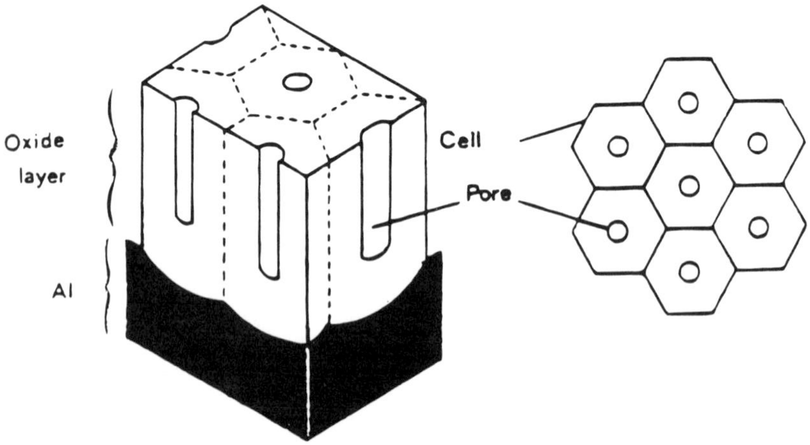

FIG. 1—*Anodized aluminum surface.*

100 to 300 Å; this corresponds to approximately 10^{12} pores per mm^2. The pores can be subsequently sealed by treating the anodized surface with hot water. The sealing is accomplished by the transformation of the anhydrous alumina into hydrated alumina (Boehmite) which is accompanied by an increase in volume [*2*]. The sealing is typically effected by immersion in boiling water in the presence of salts (acetates, oxalates). Depending on the degree of sealing, the porosity of the anodized aluminum layer can vary greatly.

The Contact Angle Technique

A drop of liquid in contact with a solid surface exhibits a contact angle. It is defined as the angle between the tangent at the drop and the substrate at the triple phase line interface (intersection point between the liquid, the solid and the gas) as shown in Fig. 2. An ideally smooth and homogeneous surface allows the drop to rest in stable equilibrium (the lowest energy state). On a rough or heterogeneous surface, the drop may reside in stable equilibrium or in one of several metastable states (energy troughs separated from neighboring states by energy barriers).

On an ideally smooth and homogeneous surface, the equilibrium contact angle is identical with the Young's contact angle γ_0. From a macroscopical point of view, addition of small volumes of liquid to the drop on such a surface will cause the drop front to advance. After the advance, the initial contact angle will be re-established. Subtraction of small volumes of liquid from the drop will cause the drop front to retract. After retraction, the initial contact angle will again be re-established.

Non idealized surfaces, such as anodized aluminum, are always rough and/or heterogenous. On such a surface, the direct determination of the equilibrium contact angle γ_0 is no longer possible by macroscopical measurement. Addition of small volumes of liquid will make the drop grow taller without moving its periphery; as a result the contact angle increases. When enough liquid is added, the drop front will suddenly advance with a leap. The angle at the onset of that advance is the maximum advancing angle noted θ_a. Removal of small volumes of liquid will flatten the drop without changing its periphery; as a result the contact angle decreases. When sufficient liquid is removed, the drop front will suddenly retract. The angle

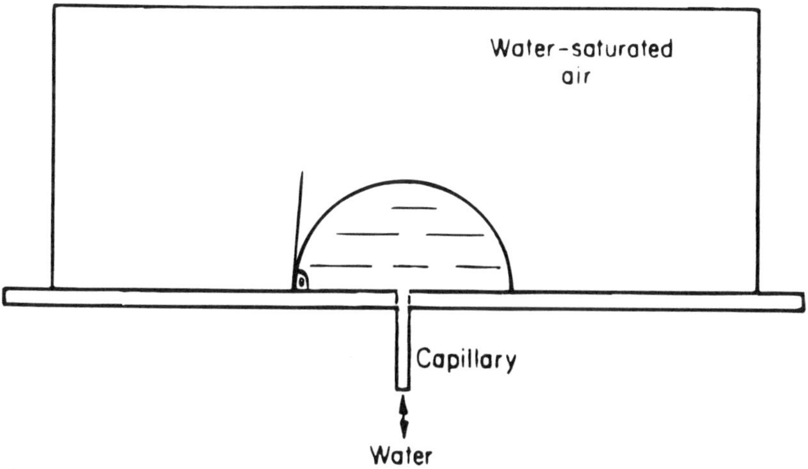

FIG. 2—*Definition of a contact angle—sessile drop method.*

at the onset of this retraction is the minimum receding contact angle noted θ_r. This phenomenon is referred to as contact angle hysteresis.

Contact angle hysteresis occurs when surface roughness or heterogeneity are on a scale larger than 0.1 μm. Hysteresis induced by surface roughness will generally not exceed 10° on polished substrates, while hysteresis on heterogeneous surfaces may be as high as 110° [*3*].

Degree of Surface Sealing (ISO 2143 Test Method)

The degree of sealing and hence the porosity of anodized aluminum substrates can be easily quantified using the ISO Test Method "Anodizing of Aluminum and its Alloys—Estimation of Loss of Absorptive Power of Anodic Oxide Coatings After Sealing—Dye Spot Test With Prior Acid Treatment" (ISO 2143). The principle of this test is to place a drop of hexafluorosilicic acid (H_2SiF_6) solution on the surface of an anodized aluminum substrate for a period of one minute. A dye solution is then applied to the surface immediately after the chemical attack. Comparing the observed color with a reference scale gives an estimate of the degree of sealing. It is obvious that this test method cannot be applied for pigmented anodized aluminum.

Electrical Impedance Measurement (ISO 2931 Test Method)

This technique is based on the analogy between an anodized aluminum surface and the equivalent electric circuit as shown in Fig. 3. Using the test method described in the ISO Test Standard "Anodizing of Aluminum and its Alloys—Assessment of Quality of Sealed Anodic Oxide Coatings by Measurement of Admittance or Impedance" (ISO 2931), the admittance of the anodized aluminum layer is measured at a frequency of 1000 Hz. This test method allows derivation of the modulus of admittance (Y) and of the phase shift (tan δ).

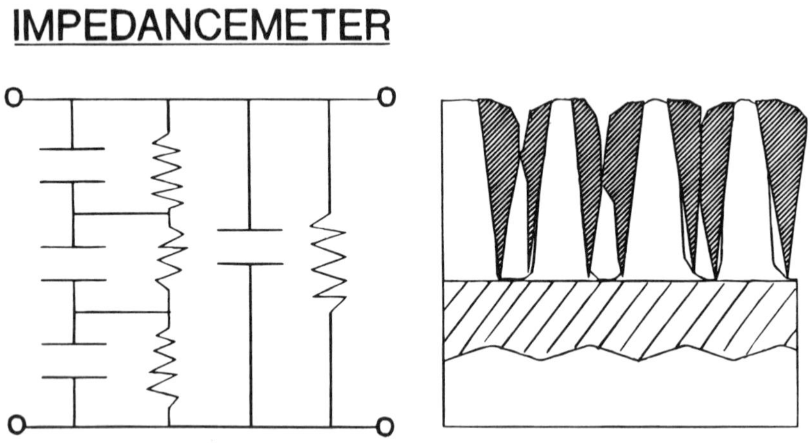

$$Z = \sqrt{R^2 + X^2}$$

$$\tan \delta = \frac{R}{X}$$

FIG. 3—*Analogy between the anodized aluminum surface and an electrical circuit.*

Since pigments are metal salts that can act as electrical conductors, the modulus of admittance is greatly affected by the nature of the pigments used to color the anodized aluminum layer. The phase shift, however, is less sensitive to the pigment nature and therefore can be used to estimate the degree of sealing.

Evaluation of Silicone Sealant Adhesion to Anodized Aluminum

Most structural silicone sealant manufacturers carry out adhesion tests to critical substrates on a routine basis as part of their project approval procedure. Dow Corning's application laboratory in Seneffe (Belgium) uses a modification of the ASTM Test Method for "Adhesion-in-Peel of Elastomeric Joint Sealants" (C 794) for evaluation of the sealant adhesion to unknown or undefined substrates. Various samples of the substrate are prepared as described in ASTM C 794 by cleaning with different solvents and, if deemed necessary, by applying primers.

After allowing sufficient time for the solvents or primers to dry, the sealant is applied as a thin layer that is reinforced by embedding a metal mesh. The samples are then allowed to cure under laboratory conditions (23±2°C) for two weeks, followed by immersion in 55°C hot, deionized water for a period of one, two, or four weeks. Directly after water immersion, the sealant is peeled away from the substrate at an angle of 180°; the peel force and type of failure being recorded. Based on the outcome of this test, recommendations on substrate preparation are made to the architect, specifier or sealant applicator.

ASTM C 794 only requires seven days water immersion at ambient temperature. The hot water immersion was chosen as a means to study the hydrolytic degradation of silicone sealant adhesion (see Fig. 4) under accelerated conditions. It can be shown that the temperature dependency of the adhesive bond hydrolysis closely follows Arrhenius' law [*4,5*]. Since the activation energy of the hydrolysis depends on the sealant formulation and the substrate, acceleration factors can only be derived for each sealant/substrate combination. For instance, for Dow Corning[R] 983 Structural Glazing Silicone Sealant applied to float glass, one week water immersion at 55°C closely reproduces the results of eight week water immersion at 23°C [*6*].

Round-robin studies conducted by members of the ASTM Committee C 24 as well as by individual members of the sealant industry indicate that ASTM C 794 peel tests made on one sealant sample with a specified substrate by a single operator in a single laboratory may yield

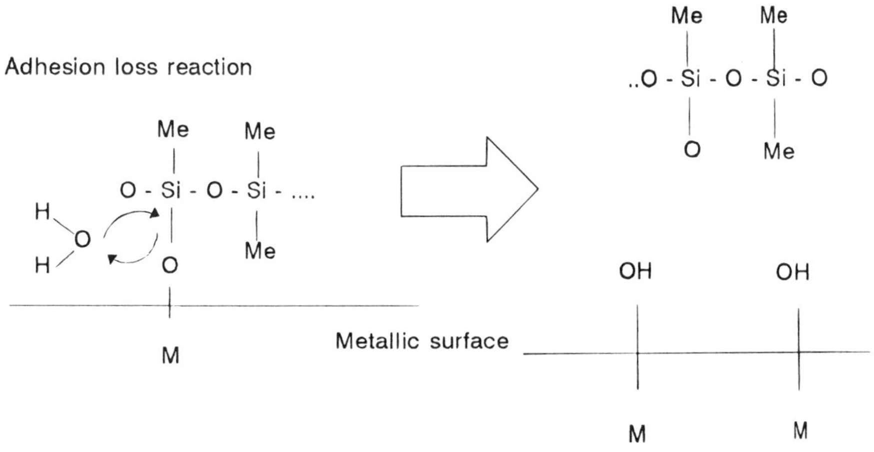

FIG. 4—*Adhesion loss mechanism due to hydrolysis of metal-siloxane bond.*

a range of values that vary by ±10 to ±20% from the mean value. It is these authors' experience that the mean peel force value averaged over triplicate samples can be reproduced with ±5 to ±10% accuracy between identical sets of triplicate samples by a single, well-trained operator within a single laboratory.

Dependency of Silicone Sealant Adhesion on the Surface Conditioning Effect of Cleaners

In order to exclude other variables from this study, only one type of anodized aluminum, supplied by Alusuisse, a company that had shown good consistency in their anodizations, was chosen to demonstrate the conditioning effect of cleaning solvents on silicone sealant adhesion. This anodized aluminum sample represented a degree of sealing close to 3 according to the ISO 2143 scale; its aluminum-oxide surface, thus, was incompletely sealed.

The anodized aluminum substrate was cleaned by wiping the surface with a solvent soaked paper tissue for approximately 10 seconds (single wipe), or by wiping the surface with a solvent soaked paper tissue followed immediately by another wipe with a different solvent (dual wipe). While the dual wipe method is generally viewed as being more effective in removing organic contaminations, most applicators favor single wipe cleaning over dual wipe cleaning because of the time saving involved.

The peel adhesion test samples were then prepared by applying Dow Corning^R 983 sealant within 5 minutes after the substrate visually appeared dry. Further peel test samples were prepared using anodized aluminum samples that had been stored in a desiccator for 72 hours following solvent cleaning. Six different commercially available cleaners were evaluated:

- A: A mild household cleaner based on ammonia, anionic and non-ionic surfactants in water (single wipe).
- B: 1,1,1-Trichloroethane followed by methylisobutylketone (CHL/MIBK) (dual wipe).
- C: Methylisobutylketone (MIBK) (single wipe).
- D: An aggressive industrial cleaner based on ammonia, anionic and non-ionic surfactants in water (single wipe).
- E: A 50/50 mixture of isopropanol and acetone (IPA/A) (single wipe).
- F: Dow Corning Q3-3556 SSG Cleaner, a proprietary siloxane based cleaner (single wipe).

The peel test samples were allowed to cure for 14 days at ambient laboratory conditions, then immersed in 55°C hot water for 7 days and immediately afterwards subjected to ASTM C 794 peel adhesion test.

Figure 5 shows the average peel strength recorded for each set of triplicate samples as a function of cleaner type and duration between cleaning and sealant application. As can be seen, peel forces measured on samples for which Dow Corning^R 983 sealant had been applied almost immediately after substrate cleaning steadily increase in the following sequence of cleaners (measurement series #1 in Fig. 5):

$$A < B < C < D < E < F$$

However, cleaners A, B, C, and E provide comparable peel forces (between 1.2 and 1.3 N/mm), when Dow Corning^R 983 sealant is applied 72 hours after cleaning the substrate (measurement series #2 in Fig. 5). Cleaners D and F provide substantially higher peel forces under the same conditions (2.3 and 2.6 N/mm, respectively).

The degree to which the various cleaners 'condition' the anodized aluminum substrate for improved silicone sealant adhesion has little relation to their cleaning effectiveness, as can be

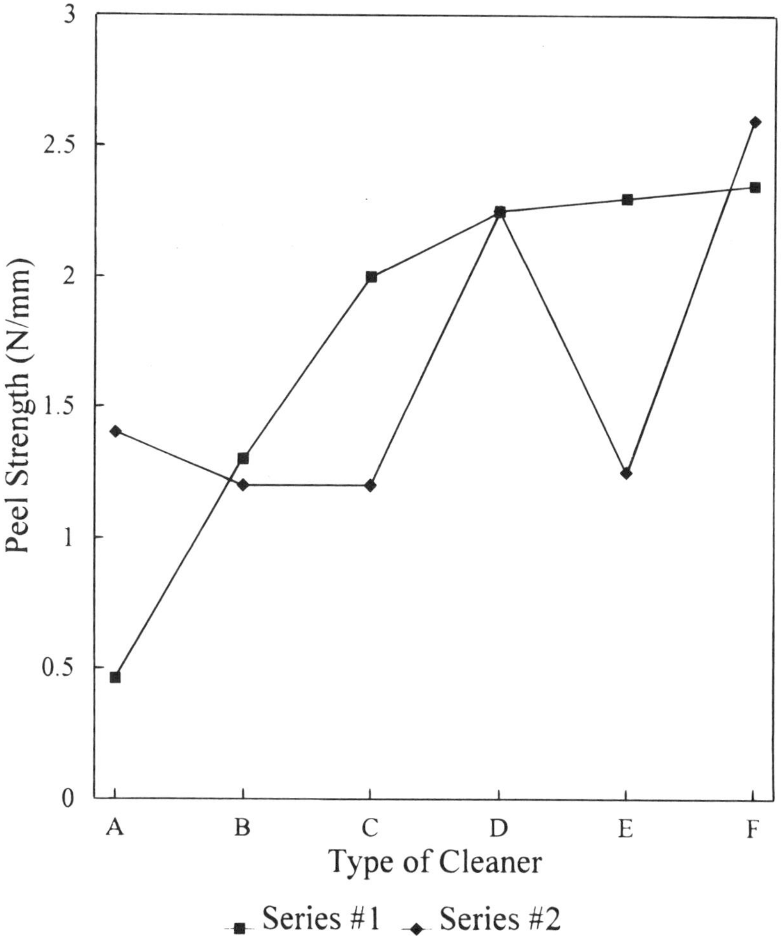

FIG. 5—*ASTM C 794 Test Method for Adhesion-in-Peel of Elastomeric Joint Sealants, peel strength values as a function of cleaner types.*

shown by studying the surface of the anodized aluminum samples as supplied and after cleaning with cleaners E and F using X-Ray Photoelectron Spectroscopy (XPS). All samples were stored for seven days in a desiccator prior to measurement to allow for solvent desorption. Overview spectra in the 0 to 1253.6 eV range were obtained for qualitative analysis, then segments of the spectra around the main peaks were recorded for semi-quantitative analysis. The XPS method provides information on the average content of these elements within the upper 5 nm thick surface layer.

The overview spectra showed the presence of carbon, oxygen, aluminum, silicon and nickel at high levels (>5%); calcium, fluorine and zinc were also detected, however, at very low levels. Based on the C_{1s} binding energy of 284.6 eV, the carbon content was attributed to the presence of a carbonaceous species with C-C and/or C-H bonds. Cleaning the anodized aluminum sample with either cleaner E or F resulted in an increase of the aluminum, nickel, calcium, fluorine and zinc signals relative to the other signals; a phenomenon which can be interpreted as a result of the decontamination of the sample surface. Al/C ratios calculated from the Al_{2p} and C_{1s} XPS peaks of the cleaned and uncleaned anodized aluminum samples show that cleaner E is far more effective in removing the organic contamination than cleaner F. Treatment of the anodized aluminum surface with cleaner E increased the Al/C ratio from 0.021 (uncleaned surface) to 0.26, while treatment with cleaner F under the same conditions only resulted in an Al/C ratio of 0.088.

It was hypothesized that by adsorbing on the anodized aluminum substrate the cleaning solvents 'condition' the surface for improved silicone sealant adhesion. It was further hypothesized that this conditioning effect diminishes, as the cleaning solvents desorb over time and a time-related re-deposition of atmospherically borne contamination occurs. In order to characterize these changes in surface properties, contact angle hystereses of deionized water on anodized aluminum were measured with a RAME-HART goniometer using the Sessile-Drop Technique as described in [7] (see Fig. 2).

Measurements of advancing and receding angles were taken two to three minutes after cleaning the anodized aluminum samples with cleaners A to F as described above and after storing the cleaned substrates for 72 hours in a desiccator. To this purpose, a hole with a diameter of 1 mm was drilled in the aluminum sample and the water was then fed through the hole using a calibrated micro-syringe. The experiments showed that the maximum advancing and the minimum receding angles as defined previously are rather difficult to measure since any external perturbation can trigger the motion of the triple line almost immediately. For that reason, the angles established immediately after the onset of the triple phase line motion were taken as the advancing and receding angles in our experiments (physical and experimental arguments which demonstrated the validity of this choice can be found in [7]). Due to the existence of polar sites at the surface of the anodized aluminum, the drops must have a diameter of at least 4 mm in order to eliminate the influence of drop size on the contact angle measurements [8,9].

The contact angle hysteresis (difference between advancing and receding angle) is plotted as a function of the cleaning method in Fig. 6. Measurement series #1 shows the contact angle hystereses observed on the anodized aluminum substrates prior to cleaning. Values of the initial contact angle hysteresis vary between approximately 17° and 21°. This variation is an indication of the accuracy of the test method and is also caused by minor changes in surface properties from one substrate sample to another. Contact angle hystereses observed immediately after substrate cleaning vary between 10° and 21° and are shown as measurement series #2. Measurement series #3 was determined after storing the cleaned substrates for 72 hours in a desiccator. As can be seen, surface properties of the anodized aluminum substrate vary greatly immediately after cleaning, however, the differences in surface properties caused by the various cleaners start to diminish over time when the cleaned substrates are stored in a desiccator.

Dependency of Silicone Sealant Adhesion on the Degree of Surface Sealing

Over a period of six months, all unpigmented anodized aluminum substrates received for project approval testing were tested for their degree of surface sealing using the ISO 2143 acid etch method prior to evaluating the adhesion of Dow Corning^R 983 sealant to the substrate. In total, 298 anodized aluminum substrates were tested. The peel test samples were prepared in triplicates as described above by wiping the anodized aluminum surface with a paper tissue soaked with Dow Corning^R Q3-3556 cleaner for approximately 10 seconds, then applying Dow Corning^R 983 sealant within 5 minutes after the substrate visually appeared dry. The samples were allowed to cure for 14 days at ambient laboratory conditions, then immersed in 55°C hot water for 7 days and immediately afterwards subjected to ASTM C 794 peel adhesion test. Figure 7 shows the correlation between Dow Corning^R 983 sealant adhesion (percentage of cohesive failure recorded) and the degree of anodized aluminum surface sealing as determined

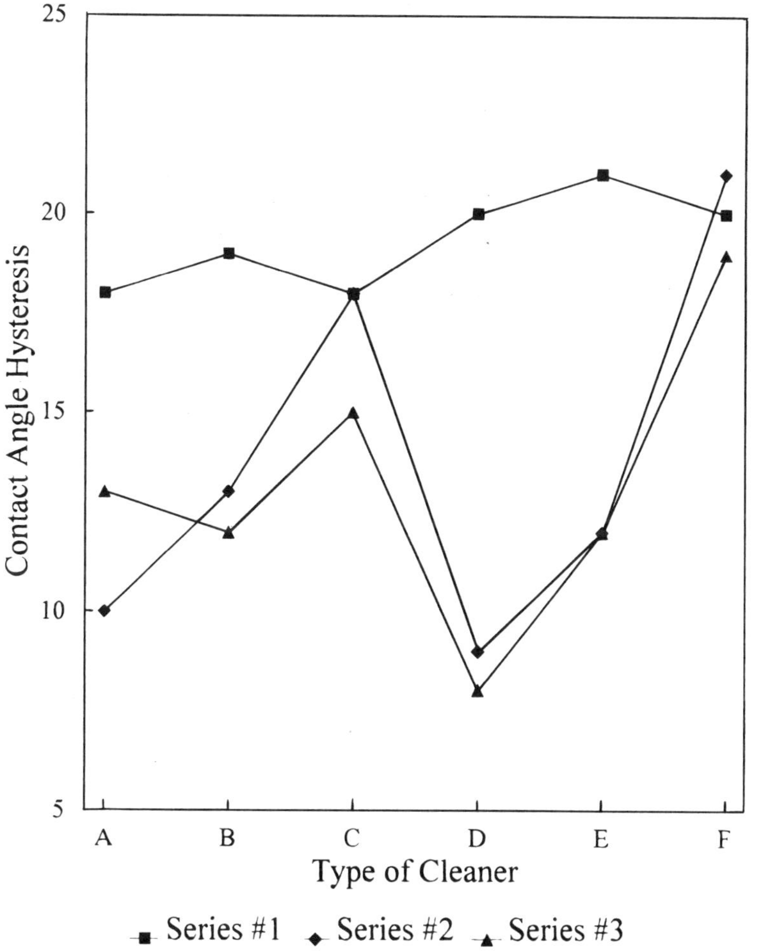

FIG. 6—*Contact angle hysteresis as a function of cleaner types.*

according to ISO 2143. The correlation gives an almost gaussian type curve with 100% of cohesive failure observed for a degree of sealing close to 1.5.

The better the surface sealing of the anodized aluminum, the smoother the substrate surface is. A smoother surface (degree of sealing: <1) results in a lower contact area between the sealant and the substrate and, thus, less mechanical interlocking and poorer adhesion. A degree of sealing of 1.5 appears to be optimum, since it provides a sufficient contact area between the sealant and the substrate, while the siloxane cleaner can more effectively penetrate into the substrate, preventing water from penetrating along the sealant/substrate interface. For a poorly sealed anodized aluminum surface (degree of sealing: >2), the roughness ratio, which is defined as the ratio between the height and the thickness of the pores, is large and the surface behaves like a perfectly smooth, nonwettable surface. This phenomenon can be attributed to the surface tension of the silicone sealant, as liquids do not penetrate pores, once the roughness ratio exceeds a certain limit [*10*].

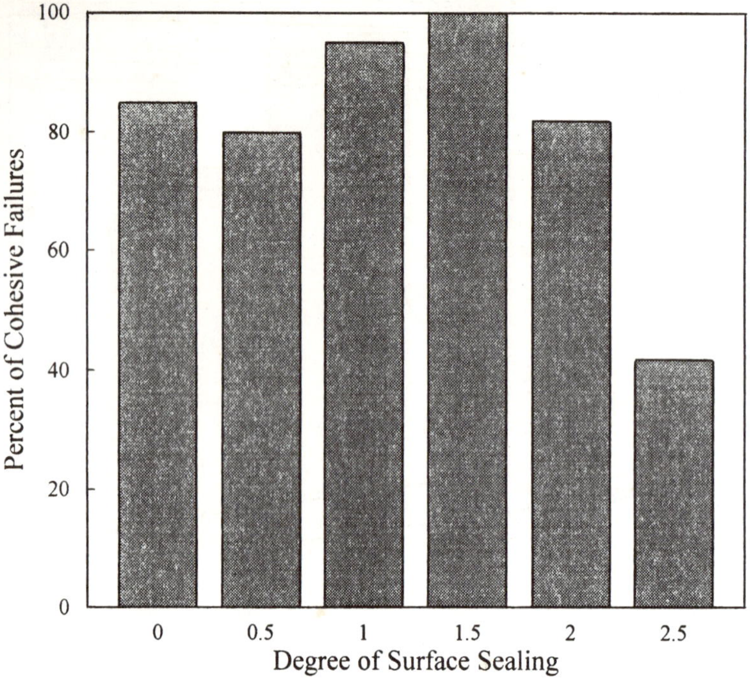

FIG. 7—*Correlation between the degree of surface sealing (measured according to ISO 2143) and the percentage of cohesive failures recorded for Dow Corning*^{R} *983 Sealant on unpigmented anodized aluminum substrates after treatment with Dow Corning*^{R} *Q3-3556 cleaner.*

Figure 8 correlates the quality of Dow Corning^{R} 983 silicone sealant adhesion in peel (ASTM C 794) with the ISO 2931 electrical phase shift measured on the anodized aluminum samples. The adhesion of Dow Corning^{R} 983 was determined after cleaning the aluminum substrates with Dow Corning^{R} Q3-3556 cleaner as described earlier. Electrical admittance tests were carried out on the anodized aluminum samples as received. In total 150 anodized aluminum samples were tested, the ratio of pigmented to unpigmented substrates being 60:40.

The quality of adhesion is expressed as the percentage of ASTM C 794 peel test samples resulting in cohesive failure mode versus the total number of samples in a given electrical phase shift class. A gaussian like correlation, similar to the one determined on unpigmented anodized aluminum substrates using the ISO 2143 etch method is observed. Based on this correlation, a tan δ between 0.4 and 0.5—this class represents approximately 85% of all anodized aluminum substrates tested—results in a cohesive failure rate close to 100%.

Methods of Predicting the Adhesion of Silicone Sealants to Anodized Aluminum Substrates

As has been shown, the quality of adhesion of silicone sealants to anodized aluminum depends on the initial surface properties of the anodized surface layer, especially its degree of sealing, the nature of the cleaner used to clean the substrate, and the time period allowed between cleaning the substrate and sealant application. For a given cleaning solvent, optimum adhesion will be observed for a certain degree of surface sealing. Measuring this degree of

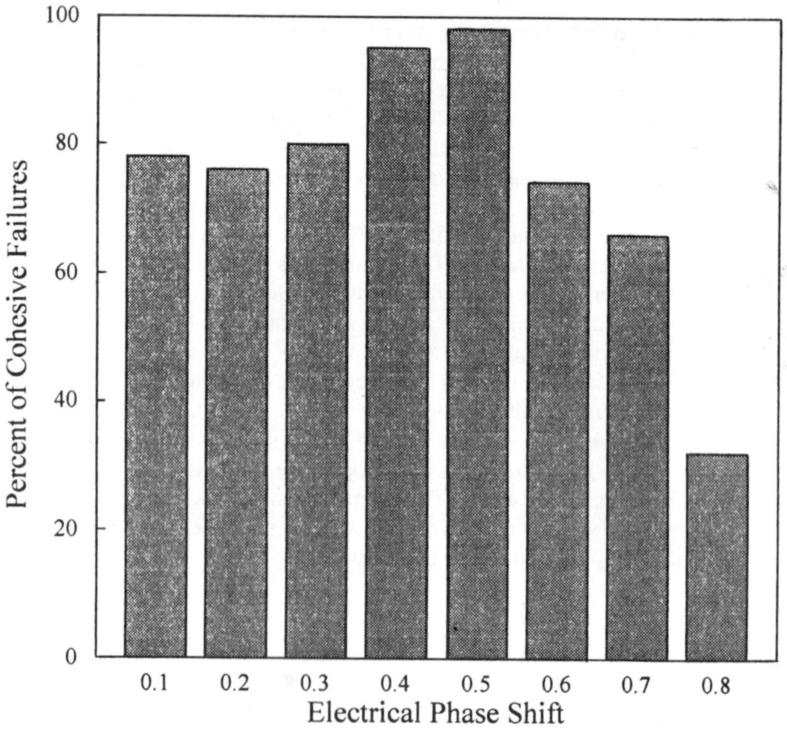

FIG. 8—*Correlation between the electrical phase shift (measured according to ISO 2931) and the percentage of cohesive failures recorded for Dow Corning*^{R} *983 sealant on pigmented and unpigmented anodized aluminum substrates after treatment with Dow Corning*^{R} *Q3-3556 cleaner.*

surface sealing allows prediction of the quality of adhesion of silicone sealants to the specific anodized aluminum substrate.

Good correlation between the lab tests and in-shop glazing results will, however, only be obtained, if the cleaner has a long-term conditioning effect on the substrate. This is the case for Dow Corning^{R} Q3-3556 SSG Cleaner.

Based on these findings, two test methods for determining the degree of sealing of anodized aluminum substrates are now being systematically used in Dow Corning's application laboratory in Seneffe in order to predict the adhesion of structural glazing sealants: The ISO 2143 test is performed on unpigmented anodized aluminum substrates, while the ISO 2931 test is used for both pigmented and unpigmented anodized aluminum substrates.

Conclusions

As has been shown above, the quality of adhesion of silicone sealants to anodized aluminum depends on the initial surface properties of the anodized surface layer, especially its degree of sealing, the nature of the cleaner used to clean the substrate, and the time period allowed between cleaning the substrate and sealant application. While the cleaners display different effectiveness in removing organic contamination from the anodized aluminum surface, this effectiveness does not correlate with the quality of the silicone sealant adhesion. More important is the 'conditioning' effect imparted by the cleaner to the anodized surface layer. The surface

conditioning effect diminishes over time; most cleaners do not condition the anodized aluminum surface for more than a few hours. Dow Corning[R] Q3-3556 SSG Cleaner provides the longest lasting conditioning effect of all cleaners considered in this study. For a given sealant/cleaner combination, optimum adhesion will be observed for a certain degree of surface sealing. Measuring the degree of surface sealing thus allows prediction of the quality of adhesion of silicone sealants to the specific anodized aluminum substrate.

Adhesion to unpigmented anodized aluminum substrates can be predicted by measuring the degree of surface sealing according to the ISO 2143 acid etch method. For pigmented anodized aluminum substrates, electrical phase shift measurements in accordance with ISO 2931 can be used to predict adhesion quality. Optimum adhesion of Dow Corning[R] 983 Structural Glazing Sealant on anodized aluminum substrates treated with Dow Corning[R] Q3-3556 SSG Cleaner is observed for a degree of sealing close to 1.5 on the ISO 2143 scale or an electrical phase shift (tan δ) between 0.4 and 0.5 measured in accordance with ISO 2931.

While this study presents results for only one silicone sealant, the underlying principles used in correlating the degree of surface sealing with the adhesion of silicone sealants on anodized aluminum substrates were found to be applicable to silicone sealants of varying cure chemistries and rheological properties. The degree of surface sealing, at which optimum adhesion is observed, however, depends on the specific sealant/cleaner combination tested. These findings will be reported in future publications.

References

[*1*] Hilliard, J. R., Parise, C. J., and Peterson, C. O. Jr., "Structural Sealant Glazing," *Sealant Technology in Glazing Systems, STP 638*, C. O. Peterson, Ed., American Society for Testing and Materials, Philadelphia, 1976, pp. 67–99.

[*2*] Lelong, P., Segond, A., and Herenguel, J., *L'Oxydation des Métaux*, Vol. 2, J. Benard, Ed., Gauthiers-Villars et Cie, Paris, 1964.

[*3*] Johnson, R. E., Jr. and Dettre, R. H., *Surface and Colloid Science*, Vol. 2, E. Matijevic, Ed., Wiley-Interscience, New York, 1969.

[*4*] McCann, K. D., "Sealant Lifetime in an Immersed Environment," Dow Corning Internal Test Report, 1990.

[*5*] Agger, J. R., Descamps, P., Iker, J., and Tilmant, G., "A Study into the Loss of Adhesion of Silicone Sealant Test Samples," Dow Corning Internal Test Report, 1991.

[*6*] Iker, J., Descamps, P., Tilmant, G., and Agger, J. R., "Etude de la Validité d'une Procédure Accélérée de Vieillissement de Pieces de Test D'Adhesivité/Cohesion par Immersion dans de L'Eau â Haute Temperature," *Revue Technique du Batiment et des Constructions Industrielles*, No. 143, March/April 1992, pp. 18–20.

[*7*] Schulze, R. D., Possart, W., Kamusewitz, H., and Bishof, C., "Young's Equilibrium Contact Angle on Rough Solid Surfaces. Part I. An Empirical Determination," *Journal of Adhesion Science and Technology*, Vol. 3, No. 1, 1989, pp. 39–48.

[*8*] Kamusewitz, H. and Possart, W., "The Static Contact Hysteresis Obtained by Different Experiments for the System PTFE/Water," *International Journal of Adhesion and Adhesives*, Vol. 5, No. 4, October 1985, pp. 211–215.

[*9*] Good, R. J. and Koo, M. N., "The Effect of Drop Size on Contact Angle," *Journal of Colloid and Interface Science*, Vol. 71, No. 2, September 1979, pp. 283–292.

[*10*] Wu, Southerry, *Polymer Interface and Adhesion*, Marcel Dekker Inc., New York, 1982.